Technical Report

Achieving Higher-Fidelity Conjunction Analyses Using Cryptography to Improve Information Sharing

Brett Hemenway, William Welser IV, Dave Baiocchi

RAND Project AIR FORCE

Prepared for the United States Air Force
Approved for public release; distribution unlimited

The research described in this report was sponsored by the United States Air Force under Contract FA7014-06-C-0001. Further information may be obtained from the Strategic Planning Division, Directorate of Plans, Hq USAF.

Library of Congress Cataloging-in-Publication Data is available for this publication.

ISBN: 978-0-8330-8166-7

RAND OFFICES

SANTA MONICA, CA • WASHINGTON, DC

PITTSBURGH, PA • NEW ORLEANS, LA • JACKSON, MS

BOSTON, MA • CAMBRIDGE, UK • BRUSSELS, BE

www.rand.org

Preface

Space debris—the man-made orbital junk that represents a collision risk to operational satellites—is a growing threat that will increasingly affect future space-related mission designs and operations. Since 2007, the number of orbiting debris objects has increased by over 40 percent as a result of the 2007 Chinese antisatellite weapon test[1] and the Iridium/Cosmos collision in 2009. With this sudden increase in debris, there is a renewed interest in reducing future debris populations using political and technical means.

The 2010 U.S. Space Policy makes several policy recommendations for addressing the space congestion problem. One of the policy's key suggestions instructs U.S. government agencies to promote the sharing of satellite positional data, as this can be used to predict (and avoid) potential collisions.[2] This type of information is referred to as space situational awareness (SSA) data, and, traditionally, it has been treated as proprietary or sensitive by the organizations that keep track of it because it could be used to reveal potential satellite vulnerabilities.

This document examines the feasibility of using modern cryptographic tools to improve SSA. Specifically, this document examines the applicability and feasibility of using cryptographically secure multiparty computation (MPC) protocols to securely compute the collision probability between two satellites. These calculations are known as conjunction analyses. MPC protocols currently exist in the cryptographic literature and would provide satellite operators with a means of computing conjunction analyses while maintaining the privacy of each operator's orbital information.

This report is written for those who are familiar with the considerations involved with data sharing as it relates to SSA. The research should be of interest to public and private sector individuals who are working on the technical and policy-oriented aspects of SSA.

This document is the latest in a series documenting RAND's research on the space debris problem. The initial monograph, *Confronting Space Debris: Strategies and Warnings from Comparable Examples Including Deepwater Horizon*, was sponsored by the Defense Advanced Research Projects Agency and was published in November 2010. The monograph addresses the debris problem by looking for applicable lessons from outside the aerospace industry. While *Confronting Space Debris* took a broad look at the debris problem, this document takes a detailed look at what steps the Air Force and other

---

[1] The Chinese test resulted in more than 35,000 pieces of debris larger than 1cm. Center for Space Standards and Innovation, *Chinese ASAT Test*, online.

[2] The White House, *National Space Policy of the United States of America*, Washington, D.C., June 28, 2010.

space-faring organizations can do to promote better situational awareness through increased data sharing.

This research was sponsored by the Director of Space Programs, Office of the Assistant Secretary of the Air Force for Acquisition. The analysis was conducted within the Force Modernization and Employment Program of RAND Project AIR FORCE as part of a fiscal year 2011 study "Space Situational Awareness and Maintaining Access and Use of Space."

RAND Project AIR FORCE

RAND Project AIR FORCE (PAF), a division of the RAND Corporation, is the U.S. Air Force's federally funded research and development center for studies and analyses. PAF provides the Air Force with independent analyses of policy alternatives affecting the development, employment, combat readiness, and support of current and future air, space, and cyber forces. Research is conducted in four programs: Force Modernization and Employment; Manpower, Personnel, and Training; Resource Management; and Strategy and Doctrine.

Additional information about PAF is available on our website:
http://www.rand.org/paf/

Contents

Figures

Tables

Summary

Using Orbital Information to Prevent Collisions in Space

The United States has been interested in protecting its on-orbit assets ever since the first U.S. satellite was launched in 1958. Since that time, the United States has been monitoring the location of objects in orbit to maintain custody of its satellite inventory, as well as predict and prevent collisions between known objects. The Space Surveillance Network (SSN), managed by U.S. Strategic Command (USSTRATCOM) and staffed by 14th Air Force, currently tracks more than 20,000 orbital objects larger than 10 cm in diameter. The data collected by the SSN are used to maintain a master catalog of known space objects, and this catalog is then used to estimate the probability of collisions involving active satellites. When a collision is predicted, the operator is notified, and evasive action can be taken.

The SSN tracks both operational and defunct satellites as well as space debris. Although debris can only be tracked passively (i.e., using sensors to detect these objects), operators of active satellites can take advantage of on-board instrumentation to provide more accurate coordinates on where the satellite is located in orbit. This operational data has an advantage over that obtained by the SSN for two reasons. First, operational data is of higher fidelity than the positional data available via the SSN. Second, the data provided by the SSN can never predict active maneuvers made by operational satellites. Any reactive tracking system like the SSN will have inherent delays in recognizing active maneuvers, and in some cases, active maneuvers can cause the SSN to temporarily lose track of the object. Both of these issues could be mitigated if satellite operators were willing to share data with each other about their satellites' positions. However, under current practice, this does not happen across the industry because the operators want to ensure that their data will be protected as private.

Privacy Concerns

The satellite community has long recognized that data sharing among operators could be used to improve space situational awareness (SSA). Although the benefits of data sharing are known, privacy concerns prevent satellite operators from sharing the accurate orbital information they possess about their satellites. Governments view such orbital information as state secrets because it could provide adversaries with insight on future intentions, and are therefore unwilling to make the information public. Private corporations view their active tracking data as proprietary information, and they fear that revealing these data would provide an advantage to their competitors.

To date, there have been some small-scale efforts to share information between operators while also providing a level of privacy. These operators employ a trusted third party to privately calculate collision probabilities for them. These calculations are called "conjunction analyses." In practice, this means that participating operators provide their private, highly accurate information to the trusted third party. The third party then performs the conjunction analyses, and returns the results to the operators. Data-sharing agreements of this type allow operators to reap the benefits of coordination and cooperation while still maintaining their own privacy. However, these data-sharing agreements require operators to find an outside party trusted by all participants. Finding such a trusted party can be difficult, and, in some cases, could be impossible, especially if the participants are large nation-states. Even if a mutually trusted party can be found, the limited availability of such trusted parties allows them to charge a premium for their services.

Privately Sharing Information Through Secure Multiparty Computation

Recent advances in the field of cryptology have produced tools that can be used to allow groups of participants to coordinate their actions, without the need for a trusted third party, while still maintaining the privacy of each individual's secrets.

These cryptographic tools are called secure multiparty computation (MPC). In its most general form, MPC allows a group of parties with private inputs to engage in a secure protocol that allows them to compute a joint function of their inputs while maintaining the privacy of each party's input.[1] In this context, an MPC protocol replaces the trusted third party, and each participant can be assured that their data will remain private, irrespective of the actions of the other participants. MPC therefore allows two operators—each with their own private orbital information—to engage in a protocol to securely compute a conjunction analysis, while maintaining the privacy of each operator's orbital information. In particular, the security of the protocol guarantees that operators learn no more than if a trusted third party had performed the conjunction analysis computation. In either setting, however, the result of the conjunction analysis computation reveals some information. For example, a conjunction analysis calculation that reports a high collision probability reveals the fact that another operator's satellite is close to your own. Whether conjunction analyses are performed by a trusted third party or via MPC, a malicious operator could attempt to learn positional information about other satellites by performing repeated conjunction analysis calculations inputting diverse (and possibly fabricated) orbits for its own satellites. The primary benefit of computing

[1] A secure protocol is a set of public rules that specify messages that each participant must send in order to execute the desired task (e.g., performing a conjunction analysis). A participant's initial message depends on his/her private inputs. Subsequent messages are crafted based on both private inputs and messages received from other participants. To be secure, the protocol must be designed so that the messages that are sent reveal nothing about the participants' inputs beyond the final result of the calculation (e.g., the collision probability).

conjunction analyses using MPC is that it allows cooperation without any need for mutual trust between the operators, because the operators cannot see each other's data.

MPC Is Feasible

The initial MPC algorithms were first developed in the 1980s;[2] since then, the cryptographic community has accepted these methods as providing mathematically provable security. Although these methods are secure, the general protocols have been seen as too inefficient for practical applications.

This report is therefore focused on determining the practical feasibility of using MPC to securely compute conjunction analyses. An MPC protocol that securely computes a conjunction analysis but requires a week's worth of computation time on modern computing hardware is of little value to the operational community. The feasibility of using MPC to compute conjunction analyses is primarily determined by the efficiency of the underlying MPC protocol.

Research Objective and Methodology

Currently, conjunction analysis calculations are never encrypted. They are performed "in the clear" by a trusted party. The research objective of this project was to determine how quickly modern MPC implementations could securely compute a conjunction analysis using present-day computing equipment. To address this task, we evaluated the efficiency of modern MPC algorithms. Efficiency depends on two factors: the complexity of the (unencrypted) conjunction analysis calculation and the efficiency of the general MPC protocol.

To determine the complexity of the conjunction analysis calculation, we reviewed the conjunction analysis literature to find a description of the algorithms in use today. We then dissected this algorithm, carefully counting the exact number of additions and multiplications needed to calculate a conjunction analysis to any specified degree of precision. To determine the efficiency of MPC protocols, we reviewed the cryptographic literature to find the benchmarks of efficiency for the most recent implementations of MPC protocols. Converting these benchmarks to "per-gate" measurements,[3] we arrived at

---

[2] Andrew C. Yao, "How to Generate and Exchange Secrets," *27th Annual Symposium on Foundations of Computer Science (FOCS 1986),* Toronto, October 27–29, 1986, pp. 162–167; Oded Goldreich, Sylvio Micali, and Avi Wigderson, "How to Play ANY Mental Game," *STOC 1987: Proceedings of the Nineteenth Annual ACM Symposium on Theory of Computing*, New York: ACM, January 1987, pp. 218–229; Oded Goldreich, *Foundations of Cryptography,* Volume II, Cambridge University Press, 2004;

[3] In digital computing, every function is represented as a binary circuit, which means that functions are broken down into a series of AND and OR gates in order to be processed by the central processing unit. Similarly, functions may be rewritten as a series of ADD and MULT gates. A circuit composed of ADD and MULT gates is called an arithmetic circuit. The time required to perform a single gate operation—

an estimate of how many milliseconds are required to perform a single addition or multiplication securely using each of the different protocols. Combining these results, we argue that securely computing conjunction analyses is a practical possibility using current algorithms and hardware.

Findings

Our research found that several MPC protocols have been developed, and many of these protocols have been implemented in software to test their efficiency. Using available benchmarks from previous MPC implementations, our estimates indicate that a number of different MPC implementations exist that could securely compute a single conjunction analysis using commercial off-the-shelf hardware in under an hour. To securely compute all conjunction analyses of interest would require computing conjunction analyses in less than ten seconds, something that is easily achieved in the insecure setting.[4] Given the rapid progress of computing hardware and the improved ease of building parallel computing systems, our findings suggest that using MPC to compute conjunction analyses is certainly possible in the coming years.[5]

Implementing Secure Multiparty Computation Protocols

As noted above, MPC eliminates the need to employ a trusted third party to perform the calculations, but a minimal amount of computer and network infrastructure is needed to enable the use of an MPC protocol. In practice, this means that each participant must have a trusted computer on which to run his or her portion of the protocol, as well as communication links between participants.

The protocol itself consists of a series of messages exchanged between the participants, at the end of which each participant learns the output of the protocol. The protocol is public, allowing each participant to independently verify that the software running on his or her own machine is valid. The MPC protocol specifies the messages that each participant must send during the execution of the protocol.

Malicious participants may be tempted to deviate from the protocol, sending malformed messages in an attempt to glean extra information about other participant's inputs. To prevent such cheating, cryptographic techniques (e.g., cut-and-choose and

---

multiplied by the number of gates needed to compute the entire function—provides a general method for estimating how long it takes to calculate any function.

[4] Hall, Robert, Salvatore Alfano, and Alan Ocampo. "Advances in Satellite Conjunction Analysis," *Proceedings of the Advanced Maui Optical and Space Surveillance Technologies Conference*, 2010

[5] It is also worth noting that our efficiency estimates are obtained by extrapolating from prior cryptographic implementations. Therefore, any MPC implementation tailored specifically for the conjunction analysis calculation would almost certainly be significantly more efficient than the algorithms we used in this analysis.

zero-knowledge proofs) exist that allow users to prove to the other participants that they are following the protocol, without revealing anything that would compromise the secrecy of their inputs.

Since each user needs only a computer (trusted by him or herself alone) and a connection to other users, MPC protocols can easily be implemented over the Internet. To calculate complex functions, the number of messages exchanged between participants can be quite large, requiring thousands of times more communication than computing the function insecurely. When the protocol requires that a large number of messages be exchanged, the data transfer speed between participants can be the performance bottleneck. When this is the case, faster, more direct data links between the participants may be necessary. A series of performance tests showed that moving from a wide area network (WAN) to local area network (LAN) yielded speedups in the range of 17–64 percent.[6]

Implications for Space Situational Awareness

Our analyses indicate that the current MPC technology is sufficiently advanced to perform secure conjunction analysis calculations quickly enough to be of use to the SSA community.

Moving forward, the next step would be to create a software prototype implementing a secure conjunction analysis calculation. Such a prototype would have a two-fold benefit. First, it would provide the most accurate running-time estimates for a real-world conjunction analysis calculation. Second, it would serve as a concrete demonstration to operators that MPC is a potentially viable means of computing conjunction analyses. Both of these effects would help the space community assess the benefits of MPC as they plan future SSA data architectures.

No matter what the efficiency or security provided by cryptographic tools, these protocols will not provide any benefit if they are not accepted by the user community. As a starting point, those operators who have already entered into data-sharing agreements are natural candidates to be the first adopters of any cryptographic secure conjunction analysis tools that might be developed based on the software prototypes.[7]

The fact that these data-sharing partnerships exist indicates a strong demand by the satellite community for high-fidelity conjunction analysis calculations on private data. The cryptographic tools discussed here have the potential to allow operators to compute high-fidelity conjunction analysis without the need for mutual trust.

---

[6] Seung Geol Choi, Kyung-Wook Hwang, Jonathan Katz, Tal Malkin, and Dan Rubenstein, *Secure Multi-Party Computation of Boolean Circuits with Applications to Privacy in On-Line Marketplaces*, IACR Cryptology ePrint Archive, Report 2011/257, 2011.

[7] For example, the operators who are part of Analytical Graphics, Inc.'s space data center, or USSTRATCOM's SSA sharing partners.

Acknowledgments

We are very grateful for our U.S. Air Force sponsors, Maj Gen John Hyten, Director, Space Programs, Office of the Assistant Secretary of the Air Force (Acquisitions, Space); and Gen C. Robert Kehler, Commander, Air Force Space Command (AFSPC/CC), who have been abundantly helpful and supportive of this work from the start.

Within RAND, Sid Dalal provided significant encouragement for this work, and we are grateful of this support, especially through the RAND Idea Showcase.

We would also like to thank the many reviewers of this work for their careful reading and insightful comments, which have greatly improved this work.

We have taken extra care to make sure that the results of this research are accessible and intuitive. The observations and conclusions made within this document are those of the authors and do not represent the official views or policies of the U.S. Air Force.

Abbreviations

AES	Advanced Encryption Standard
AGI	Analytical Graphics Inc.
BGW	Ben-Or, Goldwasser, and Wigderson
CCD	Chaum, Crépeau, and Damgård
CSSI	Center for Space Standards and Innovation
GMW	Goldreich, Micali, and Wigderson
JSpOC	Joint Space Operations Center
LAN	Local Area Network
LEO	Low Earth orbit
MPC	secure multiparty computation
NAND	"not-and"
OT	Oblivious Transfer
pdf	probability density function
SATCOM	Satellite Communications
SOCRATES	Satellite Orbital Conjunction Reports Assessing Threatening Encounters in Space
SSA	space situational awareness
SSN	Space Surveillance Network
TLE	two-line element set
USSTRATCOM	U.S. Strategic Command

1. Introduction

Since the launch of its first satellite in 1958, the United States has been interested in protecting its on-orbit assets. In order to maintain custody of its satellite inventory, and to predict and prevent collisions, the United States monitors the locations of objects in orbit. This monitoring is accomplished by the Space Surveillance Network (SSN), which is managed by U.S. Strategic Command (USSTRATCOM) and staffed by 14th Air Force. The SSN currently tracks more than 20,000 orbital objects larger than 10 cm in diameter, and the data provided by the SSN form the most important source of space situational awareness (SSA) in the world.[1] While the SSN is one of the most important sources of data concerning the locations of objects orbiting Earth, data provided by the SSN have two significant drawbacks when it comes to tracking operational satellites. First, the tracking data obtained by the SSN are significantly less accurate than the active tracking information held by each satellite's operator. Second, operational satellites can perform active maneuvers, which cannot be predicted by a passive surveillance network. This means that the SSN will have inherent delays in detecting and processing such maneuvers, which, in certain cases, may result in the SSN temporarily losing track of the object.

Operational satellites are the most important satellites to track, but the passive tracking techniques used by the SSN do not provide the most accurate positioning information. The most accurate information comes from on-board instrumentation, such as star trackers and positional gyroscopes, but this information is available only to the satellite operator. Since satellite operators maintain accurate tracking information for only their own satellites, sharing this higher-fidelity information between satellite operators could provide significantly better tracking information than what can be obtained by non-cooperative means. As an example, a comparison of cooperative and non-cooperative tracking data for Global Positioning System satellites found that cooperative tracking data reduced mean positional error by 88 percent.[2]

---

[1] Brian Weeden, Paul Cefola, and Jaganathan Sankaran, "Global Space Situational Awareness Sensors," presented at the 11th Advanced Maui Optical and Space Surveillance (AMOS) Technologies Conference, Maui, Hawaii, September 16, 2010.

[2] T. S. Kelso, David A. Vallado, Joseph Chan, and Bjorn Buckwalter, "Improved Conjunction Analysis via Collaborative Space Situational Awareness," presented at the 9th Advanced Maui Optical and Space Surveillance (AMOS) Technologies Conference, Maui, Hawaii, September 19, 2008.

Cooperative Tracking and Data Sharing Today with Trusted Providers

In 2008, a group of commercial SATCOM (satellite communications) operators maintaining satellites in the geostationary belt joined together to share data in a prototype program run by the Center for Space Standards and Innovation (CSSI), a subsidiary of Analytical Graphics, Inc. (AGI). Operators shared their private data, and CSSI's software tool SOCRATES (Satellite Orbital Conjunction Reports Assessing Threatening Encounters in Space) generated automatic notification of close approaches.[3] This service was later expanded to incorporate tracking of satellites in low Earth orbit (LEO). This system requires that all participating operators trust CSSI with their private data.

This prototype system was extended in 2010, when AGI was selected by the Space Data Association to develop and run the new Space Data Center. The Space Data Center now uses the shared (private) data to perform 300 high-accuracy conjunction analyses twice per day for objects in both geosynchronous orbit and LEO.[4] Like its predecessor, this service requires participating operators share their data with a trusted third party.

Cooperative tracking is also provided by the Joint Space Operations Center (JSpOC) under USSTRATCOM. The JSpOC uses (passively obtained) SSN data to maintain a catalog of two-line element sets (TLEs),[5] which it makes public via the Space-Track website. In addition, the JSpOC maintains a high-accuracy catalog, which is not made available publicly. The high-accuracy catalog uses information from SSA sharing program partners (who have entered into an agreement with USSTRATCOM) to provide more accurate position information for satellites operated by program partners. The high-accuracy catalog is used internally by the JSpOC to perform conjunction analyses, and satellite operators are warned of potential conjunctions involving their satellites regardless of whether they are SSA sharing program partners.

Participation in these services indicates that operators place a high value on the ability to perform conjunction analyses on high-fidelity data.

Trust and the Need for Coordination

Sharing programs like those described above require satellite operators to trust the database operator (e.g., AGI, JSpOC). This provides a significant barrier to adoption and

---

[3] Center for Space Standards and Innovation, Satellite Orbital Conjunction Reports Assessing Threatening Encounters in Space (SOCRATES), online.

[4] T. S. Kelso, "How the Space Data Center is Improving Safety of Space Operations," presented at the 13th Advanced Maui Optical and Space Surveillance (AMOS) Technologies Conference, Maui, Hawaii, September 16, 2010; Space Data Association, "Space Data Center Attains Full Operational Capability Status," press release, September 9, 2011.

[5] A two-line element set (TLE) is a data format used to convey sets of orbital elements that describe the orbits of Earth-orbiting satellites.

hence decreases the utility of these systems. Some operators are unwilling to share their data with an outside party, and those that do must pay a premium for these services.

The need for cooperation among operators and the inherent problems of mutual trust have been widely recognized in the literature.[6] Although the problems caused by a lack of data sharing between operators are well-known within the satellite community, there are currently no solutions in place that do not require operators to agree on a trusted party with whom to share their private orbital information.

Purpose and Organization of This Report

In theory, cryptographic tools such as secure multiparty computation (MPC) have the potential to improve SSA. In practice, however, implementations of these cryptographic algorithms have been too slow to be useful in their intended application.

The primary research objective of this project was to determine whether modern implementations of MPC protocols could be made fast enough to present a practical alternative for computing conjunction analyses on private data.

This report begins with an outline of the cryptographic tools known as MPC protocols, which allow stakeholders to perform functions (such as orbital conjunction analyses) that utilize inputs from each party while maintaining the secrecy of the inputs. Although MPC is not currently in use by satellite operators, it has been the subject of intense study in the cryptographic community, and general-purpose software libraries for building MPC protocols currently exist.

Chapter Two provides a technical introduction and overview of the major protocols in the cryptographic literature. Chapter Three analyzes whether MPC protocols can be made fast enough to be practical for securely computing conjunction analyses. Chapter Four summarizes the key findings and discusses how the Air Force can take steps to implement them as part of its role in preventing orbital collisions. The Appendix reviews the mathematical techniques that are used to convert a continuous integral (e.g., a conjunction analysis calculation) into an arithmetic circuit using only addition and multiplication operations.

---

[6] Jeff Foust, "A New Eye in the Sky to Keep an Eye on the Sky," *The Space Review*, May 10, 2010; Institut français des relations internationals, "Assessing the Current Dynamics of Space Security," presented at SWF-Ifri workshop, Paris, June 18–19 2009; Tiffany Chow, "SSA Sharing Program," Secure World Foundation Issue Brief, October 5, 2010; Kelso et al., 2008.

2. Overview of Secure Multiparty Computation

MPC is a cryptographic tool that allows a collection of stakeholders to compute any function of their private inputs while maintaining the secrecy of each individual's input.[1] MPC protocols allow a collection of individuals to achieve anything that could be achieved in the presence of a trusted third party, but the trusted party is replaced by a transparent and provably secure cryptographic algorithm.

Large-scale public tests of MPC protocols have been performed in the case of secure auctions, where each bidder can be sure of the privacy of his bid, yet confident that the winning bidder was chosen correctly, and in secure elections, where each vote remains private but the tally is provably correct.[2] This type of secure auction does not require a trusted auctioneer; instead, the underlying cryptographic protocol ensures the privacy of the bids as well as the integrity of the auction result. In many settings, finding a trusted party (e.g., an auctioneer or ballot counter) can be difficult or impossible, and the scarcity of trusted parties allows those that do exist to charge a premium for their services. MPC protocols have the potential to bring the benefits of cooperation and coordination of operations into realms where it was previously impossible due to lack of trust.

Since its introduction, MPC has been a subject of intense study in the cryptographic community. Surveys of the MPC literature are available from Franklin and Yung (1996), Goldreich (2004), and Lindell and Pinkas (2009).[3] The potential benefits of MPC protocols have been widely recognized, but until recently most MPC protocols were too inefficient for practical use. Recent algorithmic advances, coupled with the steady increase in computing power, are beginning to make MPC efficient enough to be practical in a wide variety of settings. Currently, general software libraries for MPC exist that provide a high-level language (similar to Java or C), and code written in this language can, in principle, be compiled into secure implementations of any desired function. The FairPlay library was an initial attempt to provide a practical

[1] The MPC protocols we consider come with rigorous mathematical proofs that guarantee the privacy of each stakeholders' input.

[2] Peter Bogetoft, Dan Lund Christensen, Ivan Damgård, et al., *Multiparty Computation Goes Live*, Report 2008/068, IACR Cryptology ePrint Archive, 2008

[3] Matthew Franklin and Moti Yung, "Varieties of Secure Distributed Computing," in *Proceedings of Sequences II, Methods in Communications, Security and Computer Science,* 1996; Oded Goldreich, *Foundations of Cryptography,* Volume II, Cambridge University Press, 2004; Yehuda Lindell and Benny Pinkas, "Secure Multiparty Computation for Privacy-Preserving Data Mining," *The Journal of Privacy and Confidentiality*, Vol. 1, No. 1, 2009, pp. 59–98.

implementation of Yao's garbled circuits (discussed below).[4] For calculations involving three or more parties, libraries also exist implementing the MPC protocols of Ben-Or, Goldwasser, and Wigderson (also discussed below).[5]

To implement an MPC protocol, it is only necessary that each participant have a trusted computer on which to run his or her portion of the protocol and a (possibly insecure) way to communicate with the other participants. The protocol consists of a series of messages exchanged between the participants, at the end of which each participant learns the output of the protocol. The protocol itself is public, allowing each participant to independently verify that the software running on his or her own machine is valid. Additional cryptographic tools can be put in place to prevent participants from deviating from the prescribed protocol. Since each user only needs a computer (trusted by him or herself alone), and a connection to other users, MPC protocols can easily be implemented over the Internet.

Privacy Concerns Arising from MPC

Before delving into the details of how MPC protocols are implemented, we briefly outline two high-level privacy concerns that are inherent in any MPC protocol. In order to prove that a protocol is secure, a *threat model* needs to be introduced that formalizes the types of attacks that could be employed against the protocol. Once an MPC protocol has been proven secure in a given threat model, users have a strong guarantee that running the protocol *leaks no more information than the output of the protocol alone*, i.e., the protocol securely simulates a trusted third party. There are many situations, however, where the output of the protocol itself may leak too much information. For example, if two satellite operators securely compute a conjunction analysis and learn that there is a high probability of collision, then even if the MPC protocol is secure, each operator will have learned a lot of information about where the other's satellite is located. In the

---

[4] Dahlia Malkhi, Noan Nisan, Benny Pinkas, and Yaron Sella, "Fairplay - A Secure Two-Party Computation System," *USENIX Security Symposium '04,* 2004; Assaf Ben-David, Noam Nisan, and Benny Pinkas, "FairplayMP—A System for Secure Multi-Party Computation," in *CCS '08 Proceedings of the 15th ACM Conference on Computer and Communications Security*, New York: ACM, 2008, pp. 257–266.

[5] Bogdanov, D., S. Laur, and J. Willemson, "Sharemind: A Framework for Fast Privacy-Preserving Computations," In *Proceedings of the 13th European Symposium on Research in Computer Security: Computer Security*, ser. ESORICS '08, Vol. 5283, Berlin, Heidelberg: Springer-Verlag, 2008, pp. 192–206.; Ivan Damgård, Martin Geisler, Mikkel Krøigaard, and Jesper B. Nielsen, "Asynchronous Multiparty Computation: Theory and Implementation," *Public Key Cryptography - PKC 2009, 12th International Conference on Practice and Theory in Public Key Cryptography, Irvine, CA, USA, March 18–20, 2009, Proceedings*, Springer, 2009, pp. 160–179; Seung Geol Choi, Kyung-Wook Hwang, Jonathan Katz, Tal Malkin, and Dan Rubenstein, *Secure Multi-Party Computation of Boolean Circuits with Applications to Privacy in On-Line Marketplaces*, IACR Cryptology ePrint Archive, Report 2011/257, 2011, pp. 416–432; Ben-David, Nisan, and Pinkas, 2008; Michael Ben-Or, Shafi Goldwasser, and Avi Wigderson, "Completeness Theorems for Non-Cryptographic Fault-Tolerant Distributed Computation," in *Proceedings of the 20th Annual ACM Symposium on Theory of Computing*, Chicago, Ill., 1988, pp. 1–10.

satellite situation, this leakage seems to be acceptable to the community, but this is a question that needs to be addressed before any MPC protocol can be securely deployed.

Privacy Threat Models

Most MPC protocols consider one of three of the following threat models describing the behavior of the participants. In order of increasingly adversarial behavior, the three models are

1. honest-but-curious (semi-honest)
2. covert
3. malicious.

In the honest-but-curious or semi-honest model, all participants are assumed to follow all protocols correctly. If the protocol dictates that the participant should send a message of a specific form, the participant will send a message of that form. In this sense, participants are assumed to be honest. On the other hand, participants are also assumed to be curious, meaning that they will attempt to analyze any information or messages that they receive to glean information about the other participants' private information.

In the covert and malicious models, participants may behave arbitrarily. In particular, they may choose not to follow the protocol, and they may send other participants malformed messages in an attempt to learn other participants' private information. The difference between the covert and malicious security models is how often a cheating participant is caught. A protocol that is covert-secure is guaranteed to detect a participant that deviates from the protocol with some fixed probability (e.g., with 75 percent probability). A protocol that is secure in the malicious model (sometimes called fully secure) will essentially always[6] detect a participant that deviates from the protocol.

The honest-but-curious setting is not intended to successfully model real-world behavior; instead, it serves as the simplest model for designing protocols. Protocols that are secure in the honest-but-curious model can often be upgraded to protocols that are secure in the covert or malicious models using standard cryptographic techniques. The honest-but-curious model thus serves as a stepping-stone, and allows protocol designers to take a more modular approach to security design.

The covert model is intended to capture situations where the penalty for cheating is high relative to the potential gain. The covert security of the protocol, coupled with the high price for cheating, serves to prevent participants from deviating from the protocol.

The fully malicious model prevents cheating entirely. Protocols that are secure in the malicious model provide the strongest security guarantees, but are the hardest to design, and consequently they are the least efficient protocols in practice.

---

[6] Formally, cheating is detected with all but negligible probability, meaning that that probability that a participant can successfully cheat approaches zero faster than the inverse of any polynomial function of the security parameter.

As this report is concerned primarily with the feasibility of using MPC for conjunction analyses, we will focus attention on the simplest threat model, the honest-but-curious setting. As MPC protocols have never been developed for performing conjunction analyses, the honest-but-curious setting provides the natural starting point for exploration. If efficient protocols can be obtained in the honest-but-curious setting, these protocols can then be adapted to obtain the security levels necessary for real-world use.

Information Leakage in MPC

MPC is designed to eliminate the need for a trusted broker without sacrificing privacy. In many situations, however, when participants work together to calculate a function based on their private information, the output of the function may reveal private information even when the *calculation* of the function does not. For example, in the conjunction analysis setting, when two satellite operators give their private orbital information to a trusted third party to compute a conjunction analysis, if the trusted party says that a collision is likely, each operator gains information about the location of the other operator's satellite. This information leakage is inherent in the conjunction analysis calculation, because it occurs when there is a trusted third party and it occurs when the trusted third party is replaced by an MPC protocol.

MPC protocols leak no more information than a trusted third party would; nevertheless, information leakage can still be a problem. For example, a satellite operator could submit the orbital information of a fleet of hypothetical satellites to the trusted party in order to learn the locations of other participants' actual satellites.

This type of attack—submitting bogus orbital information—can be discouraged by calculating only whether the collision probability is above a certain threshold, by restricting the number of conjunction analyses any operator can perform, or by comparing each operator's inputs to the computation of the low-fidelity public orbital information and issuing a warning if there is a large discrepancy.

MPC protocols are designed to mimic the functionality of a trusted third party, so any information leakage that would occur in the presence of a trusted party will also occur in the MPC protocols. While these problems are not caused by MPC, whenever MPC is implemented in a new context, potential participants must decide *whether the output of the function alone* reveals too much private information. Tools exist to help potential participants analyze the amount of information that is revealed in this way.[7]

This report is concerned with the feasibility of using MPC for conjunction analyses and does not explore the amount of information revealed by the result of a conjunction analysis. If a framework for MPC were developed for conjunction analyses, potential

---

[7] P. Mardziel, M. Hicks, J. Katz, and M. Srivatsa, "Knowledge-Oriented Secure Multiparty Computation," *Proceedings of the 7th Workshop on Programming Languages and Analysis for Security*, ser. PLAS '12. New York: ACM, 2012.

participants would need to weigh the benefits of participation against the orbital information revealed by the output of conjunction analysis.

Converting Functions into Binary Circuits

This chapter provides an overview of the two major protocols for secure two-party computation; namely, Yao's garbled circuit and the Goldreich, Micali, and Wigderson (GMW) protocol.[8] Although these protocols differ significantly, both convert the function being computed into a binary circuit, and then provide a method for securely computing each gate of the circuit using a cryptographic protocol called Oblivious Transfer (OT).

A Boolean gate is a function with a one-bit output. It will be sufficient to consider gates with two single-bit inputs and one single-bit output. One of the simplest gates is an AND gate, which outputs zero unless both input bits are one, in which case it outputs one. The "not-and" or NAND gate reverses the output of an AND gate, outputting one unless both inputs are one.

Each gate, which has two binary input wires and one binary output wire, has an associated truth table that relates the input to the output. An example is shown in Table 2.1.

Table 2.1
Truth Table for a NAND Gate

Input 1	Input 2	Output
0	0	1
0	1	1
1	0	1
1	1	0

Each truth table can be represented concisely using four bits of information, listing the outputs for each of the four possible inputs (this corresponds to the last column of Table 2.1). Simple Boolean gates can be combined to create more complex functions. Figure 2.1 shows an example of a Boolean circuit with six gates, of depth three,

[8] Andrew C. Yao, "How to Generate and Exchange Secrets," *27th Annual Symposium on Foundations of Computer Science (FOCS 1986),* Toronto, October 27–29, 1986, pp. 162–167; Oded Goldreich, Sylvio Micali, and Avi Wigderson, "How to Play ANY Mental Game," *STOC 1987: Proceedings of the Nineteenth Annual ACM Symposium on Theory of Computing*, New York: ACM, January 1987, pp. 218–229. For an in-depth discussion of these protocols, see Carmit Hazay and Yehuda Lindell, *Efficient Secure Two-Party Protocols: Techniques and Constructions*, Springer, 2010.

computing the function $[(x_1 \vee y_1) \wedge (x_2 \vee y_2)] \wedge \neg(y_2 \wedge x_3)$.[9] While any function can be represented as a circuit, computing complex functions requires extremely large circuits. For example, a circuit that computes a single multiplication of floating-point numbers requires tens of thousands of binary gates.[10]

Figure 2.1
Example of a Boolean Circuit

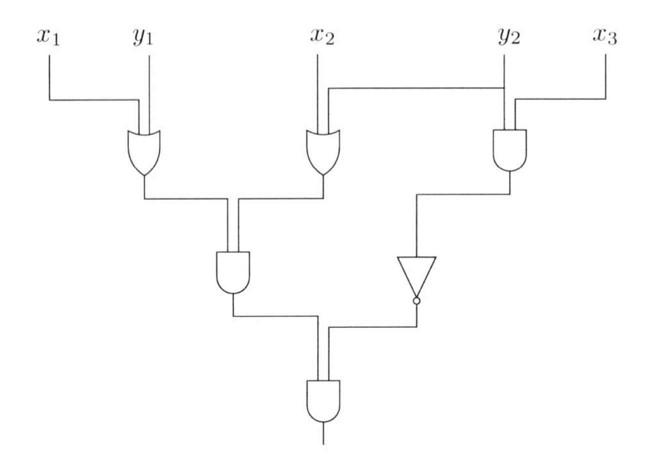

Yao's Garbled Circuit

The first protocol for secure two-party computation that this report explores is *Yao's garbled circuit*. Andrew Yao introduced the notion of MPC[11] and outlined the first two-party secure computation protocol in 1986.[12] His work described how two parties could securely calculate any public function of their joint inputs by introducing a technique that came to be known as "circuit garbling." This section provides an overview of Yao's

---

[9] The symbols ∧, ∨, and ¬ denote the binary operations AND, OR, and NOT, respectively.

[10] Reza Hashemian, "A New Multiplier Using Wallace Structure and Carry Select Adder with Pipelining," *ISCAS '02 Conference Proceedings*, 2002.

[11] Andrew C. Yao, "Protocols for Secure Computations," *23rd Annual Symposium on Foundations of Computer Science* (FOCS 1982), Chicago, Ill., November 3–5, 1982, pp. 160–164; Yao, 1986.

[12] Yao, 1986.

construction, omitting many of the technical details needed for security.[13] The work of Lindell and Pinkas provides a rigorous technical description of the protocol.[14]

Yao's garbled circuit allows two participants, denoted A and B, to compute the function $f(a,b)$, where a denotes the private input of A and b denotes the private input of B. In the case of a conjunction analysis, each party's private input is the location and velocity of their satellite and the function being calculated is the function that outputs the probability of collision.[15]

The structure of Yao's protocol is fundamentally asymmetric; one party "garbles" the circuit, and the other evaluates the garbled circuit. Despite this asymmetry in the construction, the security guarantees and outputs of the protocol can be made symmetric.

Yao's protocol (see Figure 2.2) begins as follows. Party A will garble the circuit for f gate-by-gate. To garble the gate, party A will employ a symmetric-key cryptosystem E.[16] A symmetric key cryptosystem relies on a secret key, k, and provides the guarantee that if the key is unknown the encryption $E_k(s)$ provides no information about the secret, s. Yao's technique requires two encryption steps, creating a double encryption, by encrypting the secret, s, under two different keys. This provides the guarantee that no information about the secret is leaked unless both keys are known. To garble a gate, for each input wire and each output wire of the gate, party A chooses two uniformly random keys (see Figure 2.3). Party A then creates a garbled truth table, by creating four double encryptions (requiring four secret keys), as in Table 2.2. Party A then randomly shuffles the rows of the truth table. At the end of the garbling procedure, party A has two secret keys for each wire of the circuit, and a double-encrypted truth table for each gate of the circuit.

---

[13] Yao's original protocol, as described here, only provides security against passive adversaries; Lindell and Pinkas described an extension of Yao's protocol to provide security against active adversaries. See Yehuda Lindell and Benny Pinkas, "Secure Two-Party Computation via Cut-and-Choose Oblivious Transfer," in Y. Ishai, ed., *Theory of Cryptography*, Vol. 6597 of *Lecture Notes in Computer Science*, Berlin, Heidelberg: Springer, 2011, pp. 329–346.

[14] Yehuda Lindell and Benny Pinkas, *A Proof of Security of Yao's Protocol for Two-Party Computation*, ePrint 2004/175, 2004.

[15] For example, in the case of an auction, inputs are the private values, a and b, and the function being computed is essentially the function that computes the maximum of those inputs.

[16] In practice this is a system like the Advanced Encryption Standard (AES). In the case of AES-256 a secret key is just a uniformly random 256-bit string.

Figure 2.2
Yao's Garbled Circuit Protocol

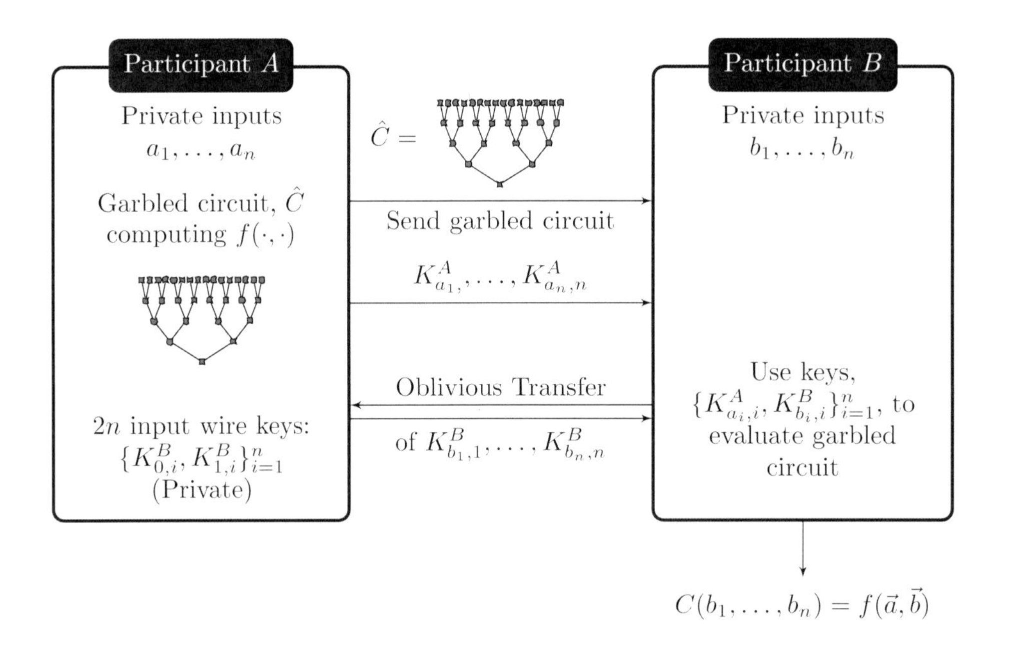

Figure 2.3
Garbled Gate

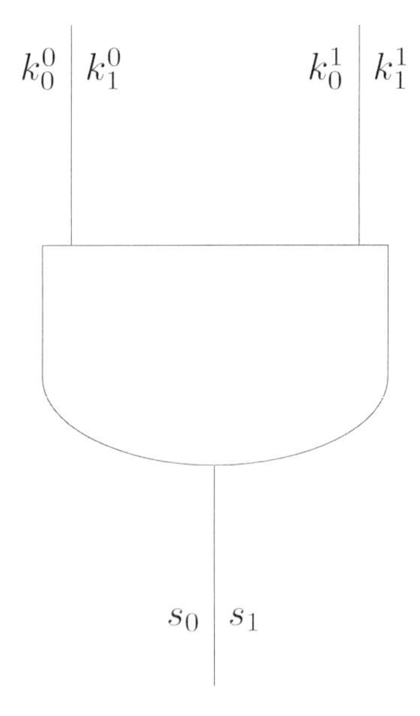

Table 2.2
Garbled Truth Table for a NAND Gate

Input 1	Input 2	Plaintext Output	Garbled Output
0	0	1	$E_{k_0^0}(E_{k_0^1}(s_1))$
0	1	1	$E_{k_0^0}(E_{k_1^1}(s_1))$
1	0	1	$E_{k_1^0}(E_{k_0^1}(s_1))$
1	1	0	$E_{k_1^0}(E_{k_1^1}(s_0))$

The important observation is that given one key from each input wire, exactly one of the four encrypted outputs can be decrypted. The values s_0 and s_1 are the two keys corresponding to the output wire of the garbled gate. The fact that the first three rows encrypt s_1 while the last row encrypts s_0 makes this a garbling of a NAND gate. In this example, given k_1^0 and k_1^1 corresponding to inputs 1, 1, the ciphertext $E_{k_1^0}(E_{k_1^1}(s_1))$ can be decrypted, and the key s_1 can be recovered (but not the fact that it corresponds to the value one). Thus having only two keys, k_1^0 and k_1^1, reveals s_1, but nothing about what type of gate was garbled. The output keys for this gate, s_0, s_1, will then be used as the keys on the input wire for the next gate in the circuit. In this way, party A will create a garbling of the entire circuit for the function f. At the end of this process, party A has created two keys for each wire and a garbled truth table for each gate. For the final gates of the circuit (the output gates), party A creates encryptions of the actual (binary) gate output instead of secret keys—i.e., for output gates $s_i = i$. Then, party A gives the entire garbled circuit (consisting of all the garbled truth tables, but not the keys) to party B. For each gate whose input wires come from party A's input, party A gives B the key to that wire corresponding to her input bit. Since this key is a uniformly random string, these keys reveal nothing about party A's input to B.

Party B now has the garbled circuit. Each wire that corresponds to one of B's inputs has two secret keys associated with it. One key is associated to an input of 0 and the other is associated to an input value of 1. To compute the circuit, party B needs the keys corresponding to his input bits on each of these input wires. At this point, party A knows both keys for each wire, but cannot reveal them both because this would reveal the entire circuit and hence A's private input. Party B knows which one out of each pair of keys that he needs, but he cannot simply reveal which key he needs because this is exactly his private input information. To allow B to acquire the necessary keys from party A, the two parties engage in an oblivious-transfer (OT) protocol (detailed below). For the input wires corresponding to party B's inputs, OT guarantees that party A does not learn party B's input bit (necessary for B's privacy) and party B learns only the key corresponding to his input bit (and not the other key). Once party B has one key (out of the pair of keys) for each input wire, he can use the garbled truth tables to compute the keys for the next level. Proceeding in this way, party B can compute the entire circuit. When B has evaluated the entire garbled circuit, he will have learned a single entry in the garbled truth tables of each output gate. By design, these values correspond to the output bits of the circuit evaluated on A and B's inputs.

This protocol requires that party A send the entire garbled circuit, and one key for each of her input wires, as well as one OT for each of party B's input bits. The remainder of the protocol does not require communication between the parties. Since Yao's original work, many significant efficiency improvements have been made.[17]

Yao's protocol only provides a method for secure two-party computation. This means that Yao's technique is not suitable for solving problems that inherently involve many parties (e.g., auctions, elections). Although there may be many satellite operators, each conjunction analysis involves only two satellites, and hence two-party computation is the appropriate model for securely computing conjunction analyses.

The Goldreich, Micali, and Wigderson Protocol

The second protocol for MPC that this report explores was developed by Goldreich, Micali, and Wigderson (called the "GMW protocol").[18] The GMW protocol provides an alternative to Yao's protocol for securely computing a function. Conceptually, the GMW protocol is very different from Yao's protocol. The GMW protocol is much more symmetric, and the underlying protocol easily extends to handle an arbitrary number of parties, whereas Yao's technique is only applicable in the two-party setting.

The GMW protocol is built on secret sharing.[19] Secret sharing is a means of distributing a secret among a number of parties, so that each party individually has no information about the secret, but together they can recover the secret.

The primary idea of the GMW protocol is to distribute each party's input using a secret-sharing scheme. Each party has a share of every party's secret input. The core of the GMW protocol is a method that allows the parties to perform a computation on the input shares in such a way that at the end of the protocol, each party is left with a share of the output.

As in Yao's protocol, the GMW protocol works on a gate-by-gate basis. The GMW protocol over the binary field proceeds as follows. Two parties, denoted A and B, with

---

[17] V. Kolesnikov and T. Schneider, "Improved Garbled Circuit: Free XOR Gates and Applications," *Proceedings of the 35th International Colloquium on Automata, Languages and Programming, Part II*, ICALP '08, Berlin, Heidelberg: Springer-Verlag, 2008, pp. 486–498; B. Pinkas, T. Schneider, N. P. Smart, and S. C. Williams, "Secure Two-Party Computation Is Practical," *Proceedings of the 15th International Conference on the Theory and Application of Cryptology and Information Security: Advances in Cryptology*, ASIACRYPT '09, Berlin, Heidelberg: Springer-Verlag, 2009, pp. 250–267..

[18] Goldreich, Micali, and Wigderson, 1987.

[19] Adi Shamir, "How to Share a Secret," *Communications of the ACM*, Vol. 22, No. 11, November 1979, pp. 612–613.

inputs a and b, begin by secret-sharing their inputs. To secretly share her input, party A chooses a random a_1, a_2 subject to the constraint $a_1 + a_2 = a$, and gives a_2 to B.[20] Similarly, party B chooses b_1, b_2 subject to the constraint that $b_1 + b_2 = b$, and gives b_1 to party A. The numbers a_1, a_2 are called shares of a. At this point A has (a_1, b_1), and B has (a_2, b_2). The share a_2 is independent of a, so B learns nothing about A's input, and vice-versa. The two parties, A and B can now perform computations on the shares as follows:

- Addition Gates: Each party starts with a share of the secret a and a share of the secret b and their goal is to end up with a share of the secret $a+b$. To do this, each party simply adds its two shares. Party A is left with $a_1 + b_1$, and party B is left with $a_2 + b_2$. Since $a+b = (a_1 + a_2) + (b_1 + b_2) = (a_1 + b_1) + (a_2 + b_2)$, each participant is left with a valid share of the sum $a+b$.
- Multiplication Gates: Each party starts with a share of the secret a and a share of the secret b and their goal is to end up with a share of the secret ab. Unlike the case of addition gates, computing multiplication gates cannot be done without communication between the participants. The goal in computing a multiplication gate, is to end up in a situation where participant A has a share c_1, and participant B has a share c_2 such that $c_1 + c_2 = ab$, and c_1, c_2 are uniformly random (but not independent). To accomplish this, party A chooses $c_1 \in \{0,1\}$ uniformly at random. The protocol will be successful if party B is left with the share $ab + c_1 = (a_1 + a_2)(b_1 + b_2) + c_1$. This is accomplished as follows. Party A computes the four values $(s_1, s_2, s_3, s_4) = (c_1 + a_1 b_1, c_1 + a_1(b_1 + 1), c_1 + (a_1 + 1)b_1, c_1 + (a_1 + 1)(b_1 + 1))$ corresponding to the four possible values of B's shares a_2, b_2. If party B can select the correct s_i corresponding to his shares, the protocol will succeed. This is accomplished using one-out-of-four OT, with party A acting as a sender with inputs (s_1, s_2, s_3, s_4), and party B acting as receiver with input $1 + 2a_2 + b_2$. This protocol allows B to learn $c_1 + (a_1 + a_2)(b_1 + b_2)$, which will be his share of the product ab.

By performing the above actions for each gate of the function being computed, the parties will end up with shares of the output of the function. These shares can then be combined to reveal the output of the function.

Like Yao's protocol, the security of the GMW protocol rests on the security of the underlying OT, and both Yao's protocol and the GMW protocol are proven to be secure, assuming the existence of a secure implementation of OT. OT (described in detail below)

[20] Throughout this section, we use arithmetic in the binary field. So addition corresponds to the XOR operation on bits and multiplication corresponds to the AND operation on bits.

is a conceptually simple cryptographic primitive, and there are many known provably secure implementations of OT, any one of which could be used to construct MPC protocols. Although both Yao's protocol and the GMW protocol rely on OT, there are some fundamental differences between the two protocols. In Yao's protocol, all of the OTs can be computed in parallel at the beginning of the communication, and the number of OTs necessary is proportional to B's input size.[21] In the GMW protocol, since each multiplication gate requires an OT, the number of rounds of communication is proportional to the depth of the circuit, and the number of OTs is proportional to the number of multiplication gates in the circuit instead of the input size.[22] Whether Yao's protocol or the GMW protocol is more efficient will depend on the type of function being evaluated.

Because Yao's protocol is a two-round protocol, and all the OTs needed in the protocol can be executed in parallel, Yao's protocol is less sensitive to the effects of network latency. The GMW protocol, on the other hand, requires a number of rounds of interaction between the participants that is proportional to the depth of the circuit being evaluated. In a multi-round protocol like GMW, the computation required in each round of the protocol cannot be started until all of the previous round's messages have been received. This means that, even if the total computation required in the GMW protocol is small, the GMW protocol may be less suitable for situations where the communication latency is high.

Oblivious Transfer

OT is a two-party protocol between a sender and a receiver, illustrated in Figure 2.4.

Figure 2.4
One-Out-of-Two OT

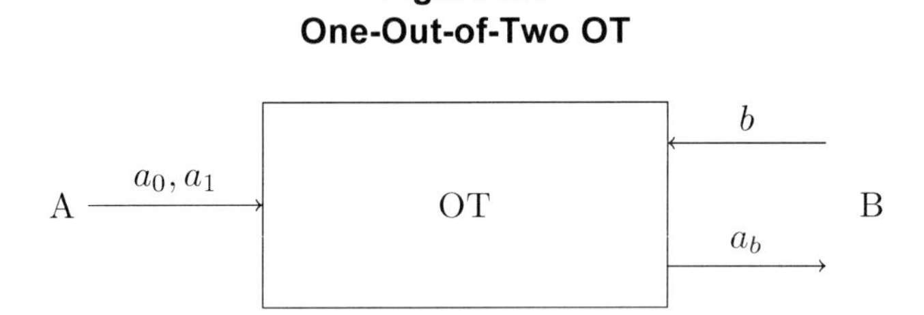

The sender, A, has inputs a_0, a_1, and the receiver, B, has a choice bit b. At the end of the protocol, B should learn the chosen input a_b. The protocol is secure if two conditions are satisfied: (1) the sender, A, does not learn the receiver's choice bit b, and

[21] Yao's protocol requires "string OTs," where the sender has two secret strings, and the receiver receives one of them.

[22] The GMW protocol uses "bit OTs," where the sender has two secret bits, and the receiver receives one of them.

(2) the receiver, B, does not learn the sender's other input a_{1-b}. Many variants of OT exist, and in particular, the GMW protocol described here requires one-out-of-four OT, where the sender has inputs, and the receiver learns one of them.

The GMW protocol requires performing OTs for each gate of the circuit being evaluated.[23] This can result in millions of OT evaluations to securely compute even fairly simple functions. Thus the efficiency of the underlying OT protocol plays a large role in determining the efficiency of the overall MPC protocol. It has been the recent improvements in the efficiency of OT that have led to major efficiency improvements in MPC.[24]

Computing an OT protocol requires two-way communication between the sender and the receiver, and requires both parties to perform (possibly expensive) cryptographic computations. To improve the online efficiency of any MPC protocol based on OT, OTs can be precomputed. Precomputing OTs is useful in scenarios where the parties know they will want to perform a calculation in the future, but do not yet know the data on which they will perform the calculation. For example, two satellite operators who know that they will want to perform a conjunction analysis tomorrow can perform all the OT calculations today. This technique cannot decrease the total amount of time necessary for the computation, but it can drastically reduce the amount of time between when the inputs are learned and when the computation finishes.

To precompute an OT, the sender picks random inputs r_0, r_1, the receiver chooses a random choice bit t, and they engage in the standard OT protocol, leaving the receiver with r_t. At a later time, to perform an OT on inputs a_0, a_1 with choice bit b, the receiver sends $t+b$, and the sender responds with $a_0 + r_{t+b}, a_1 + r_{1-(t+b)}$.[25] The receiver can recover a_b by subtracting the known value r_t. Thus, after having precomputed a random OT, any OT can be performed using only three additional bits of communication and no cryptographic calculations. By precomputing an OT, the parties can later run the OT using only three bits of communication.[26] When performing millions of OT protocols, this can result in significant savings.

---

[23] The GMW protocol also requires a number of rounds equal to the depth of the circuit. Because each round requires communication between the parties, if network latency is high, it may be prohibitively slow to have too many rounds of communication.

[24] Yuval Ishai, Joe Kilian, Kobbi Nissim, and Erez Petrank, "Extending Oblivious Transfers Efficiently," *CRYPTO*, 2003, pp. 145–161.

[25] Where all computations are done modulo 2, i.e., in the binary field where addition corresponds to the XOR operation on bits and multiplication corresponds to the AND operation on bits.

[26] Precomputing OTs can significantly reduce the amount of online computation and communication necessary in an MPC protocol, but it cannot reduce the number of rounds of communication. Thus, if the performance bottleneck is caused by network latency, precomputing OTs will have little benefit. In practice, however, it seems that computation time and network bandwidth are often the limiting factors in performance, and in these situations precomputing OTs can be beneficial.

Multiparty Computation with More Than Two Participants

MPC with more than two parties has also been studied, and many protocols have been developed, but they can all be seen as variants of the original schemes of Ben-Or, Goldwasser, and Wigderson (BGW) and Chaum, Crépeau, and Damgård (CCD).[27]

The BGW protocol can guarantee unconditional security in the case that the majority of participants follow the protocol honestly. These protocols are often called "honest majority" protocols. Unlike Yao's protocol and the GMW protocol, the BGW and CCD protocols do not use OT, and hence their efficiency is not affected by the speed of OT protocols.

In the case of a two-party computation, however, the BGW and CCD protocols cannot guarantee any type of security. A conjunction analysis is a two-party calculation, however, so the BGW and CCD protocols are not immediately applicable here. Although there may be many operators maintaining satellites, and any individual operator may wish to perform many different conjunction analyses simultaneously, but each conjunction analysis calculation remains a calculation between two parties. Thus, a two-party MPC protocol is required. There are other settings, however, where the calculations necessarily involve more than two parties. An example of a truly multiparty problem would be an auction or an election. The winner of an election, for example, cannot be computed via a series of pairwise calculations without revealing excess information. Similarly, trying to compute the highest bidder in an auction via pairwise calculations would reveal the higher bidder from every pair of bidders when only the highest bidder of the entire group needs to be revealed.

Although honest-majority MPC protocols like BGW have more limited applicability than protocols like GMW, honest-majority protocols can often be computationally more efficient than two-party protocols. To capitalize on this performance advantage, it is common to convert a two-party protocol into a three-party protocol in the following manner. If two parties, A and B wish to perform a two-party calculation, they can employ three servers S_1, S_2, S_3 and secret-share their data among the three servers. The three servers can run the secure three-party protocol and return the answer back to the original participants. This scenario is described in Figure 2.5.

---

[27] Ben-Or, Goldwasser, and Wigderson, 1988; David Chaum, Claude Crépeau, and Ivan Damgård, "Multiparty Unconditionally Secure Protocols," *Proceedings of the 20th Annual ACM Symposium on Theory of Computing*, Chicago, Ill., 1988, pp. 11–19.

Figure 2.5
Converting a Two-Party Protocol into a Three-Party Protocol

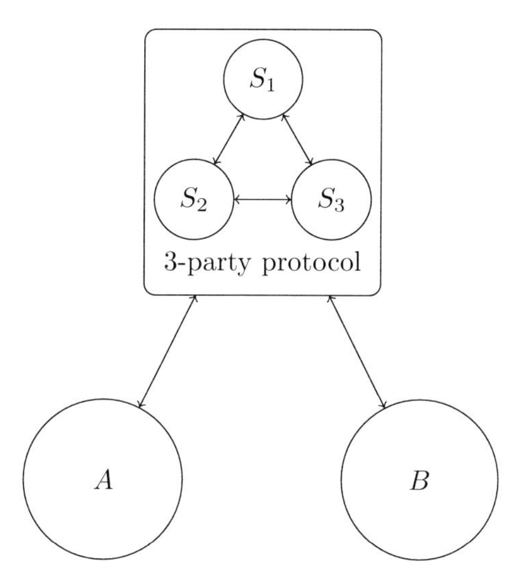

Since the three-party protocol will provide security if the majority of servers are honest, this modified protocol will provide security to the two participants if at least two out of three of the servers perform the protocol correctly. If the two parties knew exactly which server was honest, no MPC protocol would be necessary; the honest server could simply act as the trusted third party and calculate the function alone.[28]

This method of converting a three-party solution to a two-party solution can be made more computationally efficient, and hence three-party solutions have frequently been proposed for two-party problems where the two-party solution cannot be made efficient enough. The three-party solution reduces the amount of trust necessary, but does not eliminate it completely.

Requirements for Implementing MPC

Current MPC protocols are implemented as software systems. Each participant in the protocol receives an identical piece of software implementing the MPC protocol. The security of the system does not rely on any hidden aspects of the software, and hence there are no security issues involved in providing each participant with an independent copy of the software. Each client must have a trusted computer system on which to run an MPC protocol. Each client's trusted hardware will feed the client's secret inputs to the MPC software running on the system.

---

[28] Alternatively, the two parties could employ a single outside server, and the computation could be guaranteed secure as long as each party trusted the other party *or* the outside server to behave honestly.

The privacy of the protocol hinges on the fact that the randomness generated by each individual participant cannot be learned or influenced by any other participant. Generating high-quality randomness on a computer can be difficult, but it is essential to the security of the protocol, because if any participant's random choices can be predicted, the privacy of the protocol is compromised. Finally, the individual MPC software systems must be able to communicate. This is accomplished by establishing communication channels between each participant's trusted computer systems. A standard network connection suffices for this purpose, and all communication across these channels will be dictated by the MPC software implementation. The speed of the connection (both the bandwidth and the latency) will affect the speed of the MPC.

To calculate complex functions, the calculations required by each participant and the information communicated between participants can be quite large. Increased processing power will increase the speed of the local calculations, and faster data links between the participants will increase the speed of the communication. Whether the computation or communication provides the performance bottleneck will depend on the function being calculated, along with the specific computing infrastructure. In most cases, however, when performing multiparty protocols using traditional data links (e.g., the Internet) the communication provides the bottleneck and efficiency can be significantly improved by providing faster connections between the participants.

In certain situations, to maximize the speed of the network connection, it may be necessary to house each participant's trusted hardware in the same physical location. For example, a single building could be provided such that each participant has his or her own secure area where their computer systems are located. Recent tests indicate a slowdown of 17–64 percent when performing secure computations over the Internet instead of a local area network (LAN).[29]

The security of Yao's protocol and the GMW protocol rests on OT, and in the GMW protocol, performing these OTs comprises the bulk of the communication required between participants. Since OTs can be precomputed in any protocol based on OT (e.g., Yao, GMW), the participants in the MPC protocol can be constantly precomputing OT pairs. These precomputed OT pairs can then be used to securely and quickly perform the necessary secure calculations as they arise, allowing the participants to quickly perform time-sensitive calculations.

Thus far, we have discussed the protocols and requirements for MPC in general. In the following chapter, we describe how such capabilities could be applied to orbital conjunction analysis.

---

[29] Choi et al., 2011.

3. Efficiency of Implementation

From a theoretical standpoint, cryptographers have known how to compute any function, including a conjunction analysis, securely using MPC protocols like Yao's garbled circuit or the GMW protocol. The important question, however, is whether these computations can be performed efficiently enough to be of use in practice. If it takes days to perform a secure computation of a conjunction analysis, the data will be useless by the time the computation is finished. In the past, efficiency was the primary obstacle to the adoption of MPC protocols, but recent algorithmic and computational advances have improved the speed of MPC protocols to the point where they can be fast enough for many practical applications.

To answer the question of whether MPC is fast enough to be practical for conjunction analysis, we first examined the number of calculations required to perform a typical conjunction analysis. Next, we estimated the time required to perform these calculations based on benchmarks from previous implementations of MPC.

Calculating a Conjunction Analysis

This section examines the computational complexity of performing a conjunction analysis, and the time and computing resources that would be needed to perform this computation securely using a two-party MPC protocol.

A conjunction analysis is a calculation of the probability that two objects in space will collide. If the positions and velocities of the objects involved were known exactly, collisions could be predicted with certainty. In practice, error in each operator's knowledge of the positions and velocities of their satellites means that they can only estimate the probability of collision.

To perform a conjunction analysis between two objects, each operator provides their object's position, velocity, and an estimate of the error in their positional measurement. The conjunction analysis calculation then provides the probability that these two objects will collide.

Outline of Alfano's Method

This section outlines the conjunction analysis method as described by Alfano.[1] This method makes a number standard assumptions to simplify the calculations: the two objects are modeled as spheres, their relative velocity is assumed to be linear (which is

[1] Salvatore Alfano, "A Numerical Implementation of Spherical Object Collision Probability," *Journal of the Astronautical Sciences*, Vol. 53, No. 1, 2001, pp. 103–109.

approximately valid over short time intervals), and the errors in their positions are assumed to be normally and independently distributed. If the errors are normally and independently distributed, their sum is again normally distributed with variance equal to the sum of the individual variances. This observation allows all the error in position to be shifted onto one body, which simplifies the calculation. This is represented by the combined density ellipsoid in Figure 3.1.

Figure 3.1
Visualization of a Conjunction Calculation

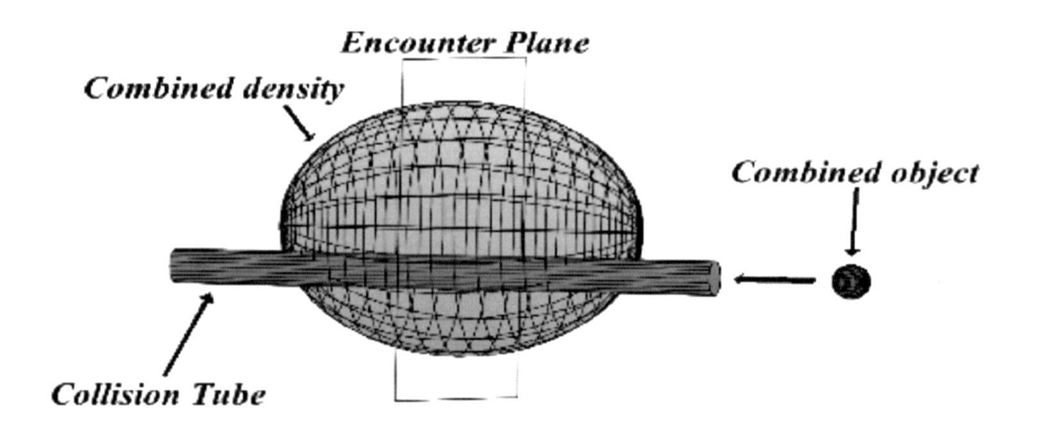

In addition to shifting the errors onto one object, it is standard to shift all the mass onto the other object, thus creating a "combined object" whose radius is equal to the sum of the radii of the two individual spheres.

The probability that the two objects collide is then exactly the same as the probability that the combined object (whose position is known exactly) collides with a point particle whose distribution is given by the combined probability density function (pdf). This probability can then be calculated as the three-dimensional integral over the path of the combined object through the combined density ellipsoid.

The assumption that the relative velocity is linear means that the integral is over a straight "collision tube." The three-dimensional integral can be reduced to a two-dimensional integral by finding the point where the combined object passes closest to the center of the combined pdf (this is where the probability of collision is highest). The plane that contains the point of closest approach and is perpendicular to the collision tube is called the "encounter plane." The three-dimensional problem is then projected onto the encounter plane, where it becomes a two-dimensional problem. The probability of collision then becomes the integral of the two-dimensional pdf in the encounter plane of the circular region defined by the cross-section of the combined object.

The two-dimensional encounter plane is represented by Figure 3.2, which shows a two-dimensional slice of the density function in Figure 3.1.

Figure 3.2
Projection onto the Encounter Plane

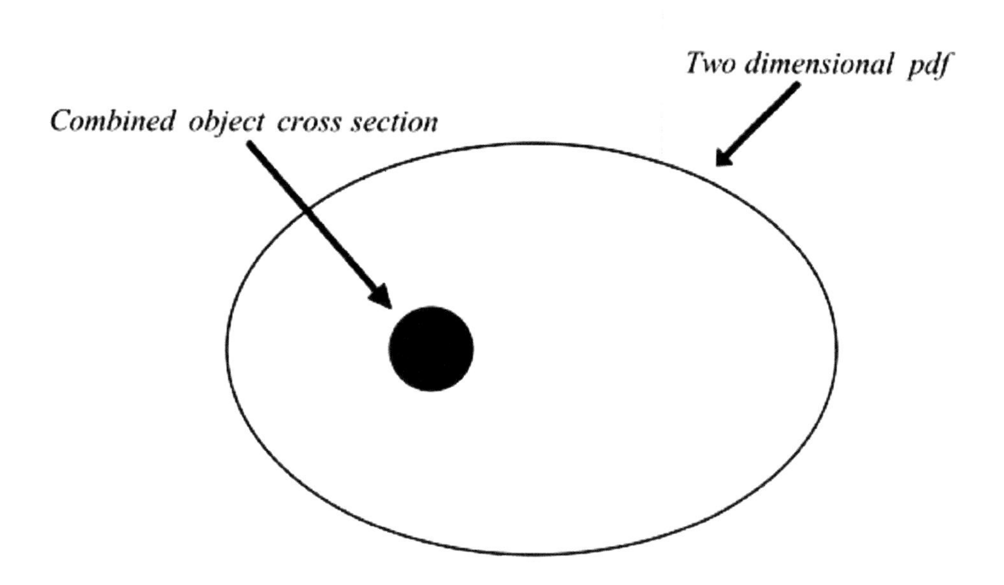

Circuit Complexity of Conjunction Analysis

Two parties who wish to compute a conjunction analysis using Alfano's method[2] perform the following computation:

- **Inputs**
  For $i \in \{1,2\}$, participant i has the following inputs:
  - A velocity vector $v_i = (v_{x_i}, v_{y_i}, v_{z_i})$, representing the velocity of participant i's object
  - A position vector $p_i = (x_i, y_i, z_i)$, representing the estimated position of participant i's object
  - An error vector $(\sigma_{x_i}, \sigma_{y_i}, \sigma_{z_i})$, representing the estimated standard deviation of the position of participant i's object. Let $C_i = \begin{pmatrix} \sigma_{x_i} & 0 & 0 \\ 0 & \sigma_{y_i} & 0 \\ 0 & 0 & \sigma_{z_i} \end{pmatrix}$
  - A radius R_i, representing the radius of participant i's object.

---

[2] Alfano, 2001.

- **Step 1**: Calculating the Encounter Plane:
  The parties compute the relative velocity as $r_v = (v_{x_2}, v_{y_2}, v_{z_2}) - (v_{x_1}, v_{y_1}, v_{z_1})$.
  The encounter plane is perpendicular to the relative velocity, r_v, so an orthonormal basis for the encounter plane can be computed by setting:

$$\vec{i} = \frac{r_v}{\|r_v\|}, \vec{j} = \frac{v_2 \times v_1}{\|v_2 \times v_1\|}, \vec{k} = \vec{i} \times \vec{j}$$

  Next, this basis needs to be rotated so that it is parallel to the principle axes of the projected ellipse. This can be done as follows. Let Q be the three-by-two matrix with columns $\vec{j}, \vec{k}$, then set $C = Q^T(C_1 + C_2)Q$. Thus, C is a two-by-two matrix. Let u, v be the normalized eigenvectors of C and let σ_x, σ_y be the square roots of their eigenvalues. The vectors u, v form an orthonormal basis for the encounter plane, and so it only remains to project the relative position into the u, v coordinate system. This can be done by setting:

$$\begin{pmatrix} x_m \\ y_m \end{pmatrix} = \begin{pmatrix} u & v \end{pmatrix}^T Q^T (p_2 - p_1)$$

  Finally they calculate the combined radius, R, which is the sum of the radii of the two objects, $R = R_1 + R_2$.

- **Step 2**: Calculation of Probability:
  Now that all the information has been shifted to the encounter plane, the two parties can calculate the probability of collision, P. The probability of collision is given by the two dimensional integral

$$P = \frac{1}{2\pi\sigma_x\sigma_y} \int_{-R}^{R} \int_{-\sqrt{R^2-x^2}}^{\sqrt{R^2-x^2}} \exp\left[\frac{-1}{2}\left[\left(\frac{x - x_m}{\sigma_x}\right)^2 + \left(\frac{y - y_m}{\sigma_y}\right)^2\right]\right] dydx$$

  By standard techniques, this can be converted into a single integral

$$P = \frac{1}{\sqrt{8\pi}\sigma_x} \int_{-R}^{R} \left[\operatorname{erf}\left(\frac{y_m + \sqrt{R^2 - x^2}}{\sqrt{2}\sigma_y}\right) + \operatorname{erf}\left(\frac{-y_m + \sqrt{R^2 - x^2}}{\sqrt{2}\sigma_y}\right)\right] \exp\left(\frac{-(x + x_m)^2}{2\sigma_x^2}\right) dx$$

  where

$$\operatorname{erf}(z) = \frac{2}{\sqrt{\pi}} \int_0^z e^{-t^2} dt$$

  is the standard error function.[3] Approximating the integral by a Riemann sum[4] yields the following:

[3] See the "Error Function" subsection of the Appendix for a review of the error function.

[4] See the "Estimating Integrals" subsection of the Appendix for information on the Riemann sum.

$$P \approx \frac{R}{\sqrt{8\pi}\sigma_x n} \sum_{i=0}^{n} \left[\operatorname{erf}\left(\frac{y_m + \frac{2R}{n}\sqrt{(n-i)i}}{\sqrt{2}\sigma_y} \right) + \operatorname{erf}\left(\frac{-y_m + \frac{2R}{n}\sqrt{(n-i)i}}{\sqrt{2}\sigma_y} \right) \right] \exp\left(\frac{-\left(\frac{R(2i-n)}{n} + x_m \right)^2}{2\sigma_x^2} \right).$$

Alfano suggests using Simpson's Rule to obtain a more accurate approximation than a simple Riemann summation.[5] This method is described and analyzed below. Using the symmetry of the integrand on the domain [-R,R], the integral simplifies to

$$P = \frac{1}{\sqrt{8\pi}\sigma_x} \int_0^R \left[\operatorname{erf}\left(\frac{y_m + \sqrt{R^2 - x^2}}{\sqrt{2}\sigma_y} \right) + \operatorname{erf}\left(\frac{-y_m + \sqrt{R^2 - x^2}}{\sqrt{2}\sigma_y} \right) \right] \left[\exp\left(\frac{-(x + x_m)^2}{2\sigma_x^2} \right) + \exp\left(\frac{-(-x + x_m)^2}{2\sigma_x^2} \right) \right] dx$$

The general form of Simpson's rule yields an approximation of the form[6]

$$\int_0^R f(x)dx \approx \frac{\Delta x}{3} \left(f(0) + f(R) + \sum_{i=1}^{n} 4f(x_{2i-1}) + \sum_{i=1}^{n-1} 2f(x_{2i}) \right)$$

where $\Delta x = R/2n$, and $x_i = i\Delta x$. In a conjunction analysis, the integrand, $f(x)$, is

$$f(x) = \left[\operatorname{erf}\left(\frac{y_m + \sqrt{R^2 - x^2}}{\sqrt{2}\sigma_y} \right) + \operatorname{erf}\left(\frac{-y_m + \sqrt{R^2 - x^2}}{\sqrt{2}\sigma_y} \right) \right] \left[\exp\left(\frac{-(x + x_m)^2}{2\sigma_x^2} \right) + \exp\left(\frac{-(-x + x_m)^2}{2\sigma_x^2} \right) \right]$$

Approximating the integral using Simpson's rule with n terms can thus be done with $2n + 1$ parallel evaluations of the integrand f, followed by $2n$ additions. To evaluate the integrand, $f(x)$, both erf and exp will be approximated by a series expansion.[7]

We can use the expansion

$$\operatorname{erf} x = \frac{2}{\sqrt{\pi}} \sum_{n=0}^{\infty} \frac{(-1)^n x^{2n+1}}{n!(2n+1)}, \qquad \exp x = \sum_{n=0}^{\infty} \frac{x^n}{n!}.$$

---

[5] Alfano, 2001.

[6] See the "Simpson's Rule" subsection of the Appendix for a review of Simpson's rule.

[7] See the "Evaluating Functions" subsection of the Appendix for a review of the Taylor series approximation.

Truncating the expansion yields an error,[8]

$$\left|\operatorname{erf} x - \frac{2}{\sqrt{\pi}}\sum_{n=0}^{N}\frac{(-1)^n x^{2n+1}}{n!(2n+1)}\right| \le \frac{2}{\sqrt{\pi}}\frac{x^{2N+1}}{N!(2N+1)}.$$

Calculating the sum

$$\sum_{n=0}^{N}\frac{(-1)^n x^{2n+1}}{n!(2n+1)}$$

for some shared x requires only N multiplications to generate the sequence $x, x^3, \ldots, x^{2N+1}$. The other values are constants, and the sum can be done locally. Similarly, the N term Taylor expansion for $\exp$ can be evaluated with N multiplications. Each evaluation of $f(x)$ requires two evaluations of erf and two evaluations of $\exp$, followed by a single distributed multiplication of the results.

Approximating erf to N_1 terms and exp to N_2 terms, requires $2N_1$ distributed multiplications to calculate the erf, $2N_2$ distributed multiplications to calculate the $\exp$, and one to multiply them. This is done $2n+1$ times (once for each term in Simpson's approximation), yielding $(2n+1)(2N_1+2N_2+1)$ multiplications. Finally, two more distributed multiplications are needed to incorporate the leading terms R and $\frac{1}{\sigma_x}$. This yields a total of $(2n+1)(2N_1+2N_2+1)+2$ distributed multiplications.

In this analysis, we assume that each object's position, velocity, error vector, and radius must remain private. If some values could be shared, the performance of the secure computations could be significantly improved.

Estimating the complexity of computing this function securely requires counting the number of additions and multiplications in the computation, as well as estimating the number of rounds of communication necessary to compute the function using a protocol like GMW or a three-party protocol like BGW or CCD.[9] In these protocols, the number of rounds corresponds to the depth of the circuit computing the conjunction analysis. On the other hand, an implementation using Yao's garbled circuit can always be done with a single round of communication.

---

[8] Sylvain Chevillard and Nathalie Revol, "Computation of the Error Function erf in Arbitrary Precision with Correct Rounding," *RNC 8, the 8th Conference on Real Numbers and Computers*, Santiago de Compostela, Spain, 2008, pp. 27–36.

[9] In these protocols, the number of rounds is essentially the depth of the circuit computing the conjunction analysis.

- Circuit Complexity of Calculating the Encounter Plane:
The complexity of the conjunction analysis calculation will be dominated by the complexity of the calculation of the integral, and the precise number of gates needed to calculate the encounter plane is unimportant. To make the calculation of the encounter plane feasible, however, it is important to note that the eigenvectors and eigenvalues of a two-by-two symmetric matrix can be calculated by a closed-form equation involving addition, multiplication, and a square root.
- Circuit Complexity of Probability Calculation:
  - Calculating the leading term $\frac{R}{\sqrt{8\pi}\sigma_x n}$ requires 1 multiplication.
  - We can evaluate each summand in parallel, so we examine the complexity of calculating each summand.

    The two erf s, and the two exp s can be computed in parallel. Calculating the argument for erf requires 1 multiplication and 1 addition, and calculating a series approximation to N_1 terms then requires N_1 sequential multiplications. Since we must calculate erf at two points, this is $2(N_1+1)$ multiplications and two additions in N_1+2 rounds. Calculating exp s using an N_2 term approximation requires one addition and one multiplication to compute the argument, followed by N_2 multiplications. Since we must compute two in parallel, this is $2N_2$ multiplications in N_2 rounds. Adding the two erf s and the two exp s and multiplying requires two more additions and one multiplication and adds one round. This gives a total of $2N_1 + 2N_2 + 1$ multiplications and four additions in $\max(N_1+3, N_2+2)$ rounds.

Since there are $2n + 1$ terms in the sum, putting it all together we have $(2n + 1)(2N_1 + 2N_2 + 1)$ multiplications and $5n$ additions in $\max(N_1+4, N_2+3)$ rounds. When erf is calculated with an N_1 term approximation, exp is calculated with an N_2 term approximation, and Simpson's rule is carried to n terms, the total number of multiplications is summarized in Table 3.1.

Table 3.1
Circuit Complexity of Integration

Step	Multiplications	Additions	Rounds
$\frac{R}{\sqrt{8\pi}\sigma_x n}$	1	0	1
Each Summand	$2N_1 + 2N_2 + 1$	4	$\max(N_1+3, N_2+2)$
Total	$(2n+1)(2N_1+2N_2+1)$	$5n$	$\max(N_1+4, N_2+3)$

As the values for n, N_1, and N_2 increase, the calculation becomes more accurate, and the running time of the conjunction analysis increases, as shown in Table 3.1. The values of n, N_1, and N_2 do not change the privacy, only the accuracy, of the final result, and this trade-off between running time and precision exists when computing conjunction analyses even without any privacy concerns. The values for n, N_1, and N_2 that yield sufficiently accurate estimates of the probability of conjunction can be determined empirically. Alfano suggests that n need be no larger than 50.[10]

Time Estimates

Based on the preceding analysis, we next made rough estimates of the amount of time it takes to compute a conjunction analysis securely. The estimates are rough because the time required is governed by a number of factors, the most important of which are

- the desired numerical precision of the calculation
- the processing speed of each participant's hardware system
- the bandwidth of the connection between participants
- the latency of the connection between participants.

The calculations in the preceding section show the trade-off between the desired precision of the calculation and the number of local arithmetic operations required by each participant and the number of rounds in the protocol. The amount of time required to perform the local arithmetic operations is dependent on the processing speed of each participant's hardware system. When performing these conjunction analyses, the primary bottleneck may be the latency of the connection between the participants, and not the processing power of each participant. In general, the number of rounds needed to compute a function using the GMW, BGW, or CCD protocols is proportional to the depth of the circuit computing the function. An exception is Yao's garbled circuit, which can always be performed in one round. As discussed above, each round involves sending messages between the participants, and hence each round can proceed no faster than the latency of the network connection.

One of the difficulties in a theoretical analysis of the complexity of an MPC calculation comes from determining whether computation or communication will provide the performance bottleneck. The relative speeds of computation and communication are highly dependent on the actual hardware architecture and software being employed. It is useful, therefore, to determine these values empirically. The dependency on latency was highlighted in recent practical tests of MPC protocols, which showed that MPC protocols run 17–64 percent slower when participants are connected via the Internet instead of a local area network (LAN). This gives an estimate of the performance benefits that could be obtained by locating each operator's server in the same physical location.

[10] Alfano, 2001.

The most effective method for determining the performance characteristics of an MPC would be to implement and benchmark secure conjunction analysis protocols. One reason for this is that the running time of an MPC protocol is highly dependent on the structure of the circuit being computed. Beyond this, a single function may have many different circuit representations, and choosing which one to use may depend on the MPC protocol being employed. For example, when using the GMW protocol in an environment with high network latency, minimizing the depth of the circuit might be most important. On the other hand, when using Yao's protocol, minimizing the number of multiplication gates might have a bigger performance benefit than minimizing the depth.

Conclusion

The analysis in the preceding section reveals that calculating a conjunction analysis takes about $2n(N_1 + N_2)$ floating point multiplications. Using Alfano's bound that 50 terms in Simpson's Rule suffices ($n = 50$), and using 50 terms in the expansions of erf and exp ($N_1 = N_2 = 50$), yields a calculation requiring approximately 10,000 secure floating point multiplications. Using the estimate that a single floating point multiplication requires approximately 10,000 binary gates,[11] calculating a conjunction analysis to high precision will require approximately 100 million binary gates.

While the conjunction analysis circuit is large, the size of the conjunction analysis circuit does not put it beyond the reach of existing MPC technology. In fact, implementations of Yao's garbled circuit have been tested on circuits containing millions[12] and even billions of gates.[13] Even in the malicious model, billion-gate secure two-party computations have been demonstrated using Yao's protocol.[14] The work of Huang et al.[15] reports evaluating a circuit with 1.29 billion gates in 223 minutes, using commercial off-the-shelf hardware. Extrapolating down, this suggests that evaluating a conjunction analysis circuit of about 100 million gates should take about 17 minutes. Conjunction analyses can be computed for events that are up to a few days in the future. Thus, a computation time of tens of minutes for a single conjunction analysis would

---

[11] Reza Hashemian, 2002.

[12] Lior Malka, "VMCrypt: Modular Software Architecture for Scalable Secure Computation," *Proceedings of the 18th ACM Conference on Computer and Communications Security*, ser. CCS '11. New York: ACM, 2011, pp. 715–724.

[13] Y. Huang, C. H. Shen, D. Evans, J. Katz, and A. Shelat, "Efficient Secure Computation with Garbled Circuits," *Proceedings of the 7th International Conference on Information Systems Security*, ser. ICISS'11. Berlin, Heidelberg: Springer-Verlag, 2011, pp. 28–48;

[14] B, Kreuter, Abhi Shelat, and Chi-Hao Shen, "Billion-Gate Secure Computation with Malicious Adversaries," *Proceedings of the 21st USENIX Conference on Security Symposium*, ser. Security'12, Berkeley, Calif.: USENIX Association, 2012, p. 14.

[15] Huang et al, 2011.

allow a single machine to compute hundreds of conjunction analyses in a short enough time frame to allow for early warning of any forecasted conjunction events.

This estimate is extremely rough, but it provides an indication that computations requiring millions of binary gates (such as conjunction analysis) could be performed on the order of minutes using existing implementations of MPC and off-the-shelf hardware. It is also important to note that this estimate is more likely to be high than low, for several reasons. First, the circuit computing an insecure conjunction analysis can almost certainly be modified to make it more amenable to computation using different MPC protocols. Second, the underlying cryptographic algorithms are in a period of rapid advancement, and the general efficiency of the protocols is improving. Third, hardware and network speeds are also improving at a rapid rate. The next step would be to obtain more-accurate time estimates by actually implementing and testing a prototypical conjunction analysis using an existing MPC software framework.

Securely computing a conjunction will always be somewhat slower than performing the same computation without security considerations,[16] but not every conjunction analysis needs to be computed in a secure manner. Conjunction analyses performed using the public catalogs can be used as a filter. Secure conjunction analyses can be restricted to those objects that are found to have some threshold probability of collision using the publicly available data. This type of filtering, using rough estimates of collision probability to determine which objects need more accurate scrutiny, is already in place, and could easily be adapted to the MPC setting.

[16] Using the software package Maple, computing a conjunction analysis with no privacy considerations takes less than a second on a standard desktop computer.

4. Conclusions and Recommendations

This research addressed the question of whether modern cryptographic tools can be used to improve SSA by facilitating secure conjunction analysis calculations without requiring operators to reveal their private orbital information to any outside party. Many cryptographic tools have been developed that allow multiple participants to engage in arbitrary secure computations. To be of value, however, a cryptographic protocol must be both secure and efficient enough to use in practice. Our analysis has focused on the efficiency of these protocols because the security of the underlying cryptographic algorithms has been rigorously and mathematically proven in the cryptographic literature.

Our analysis indicates that the complexity of the conjunction analysis calculations is low enough to be computed securely using existing MPC algorithms. The next step would be to develop a software prototype implementing a secure version of the conjunction analysis calculation. This prototype would provide the most effective means of accurately determining the running time (and hardware requirements) of a secure conjunction analysis calculation. Such a prototype could also serve as a demonstration to the community that MPC tools provide an effective method for securely computing conjunction analyses.

Appendix: Mathematical Background

Introduction

MPC protocols, like those based on the GMW protocol or Yao's garbled circuit, provide a means for performing any computation securely. To achieve this, these MPC protocols provide a means for computing individual gates (e.g., ADD, MULT, AND, OR, NAND) securely. These secure gates can then be composed to compute a larger circuit securely.

Cryptographic constructions like those of Yao and GMW can be seen as compilers that take a public circuit that implements a desired functionality, and compile it into a secure circuit that computes the same functionality. The cryptographic literature does not address how to transform a desired function (e.g., the conjunction analysis integrals) into a (public) circuit that computes the same functionality using only addition and multiplication operations. This transformation instead relies on basic tools from numerical analysis.

This appendix reviews the mathematical techniques that are used to convert a continuous integral (e.g., a conjunction analysis calculation) into an arithmetic circuit using only addition and multiplication operations.

Evaluating Functions

A polynomial is a function of the form $f(x) = a_n x^n + a_{n-1} x^{n-1} + \cdots + a_1 x + a_0$.

Polynomials are some of the most well-behaved functions in mathematics, and, consequently, working with polynomials has many advantages over working with arbitrary functions. An important property of polynomials is that they can be evaluated using only addition and multiplication.

For example, evaluating a polynomial, $f(x)$, of degree n (as above) requires $n-1$ multiplications to compute the series $x, x^2, x^3, \ldots, x^n$; n more multiplications to compute the monomials $a_n x^n, a_{n-1} x^{n-1}, \ldots, a_1 x$; and finally n additions to add the $n+1$ terms together. Thus, evaluating f at any point can be done using $2n-1$ multiplications and n additions.

While evaluating a polynomial can be done efficiently using only addition and multiplication, evaluating general functions can be difficult. This difficulty means that, in practice, it is often necessary to approximate a more complicated function by a polynomial. Such an approximation provides a method for evaluating complicated functions using only the operations of addition and multiplication.

One of the most general and effective methods for approximating a function by a polynomial is called the Taylor Series approximation, introduced by Brook Taylor in 1715. Given a function f(x), the number $f^{'}(a)$ represents the slope of the tangent line to f at the point a. The equation of the tangent line is given by $l_a(x) = f(a) + f^{'}(a)(x-a)$.

Figure A.1 gives an example of the tangent line approximation. The tangent line, $l_a(x)$ provides the best linear approximation to the function at the point a.

Figure A.1
Approximation by a Tangent Line

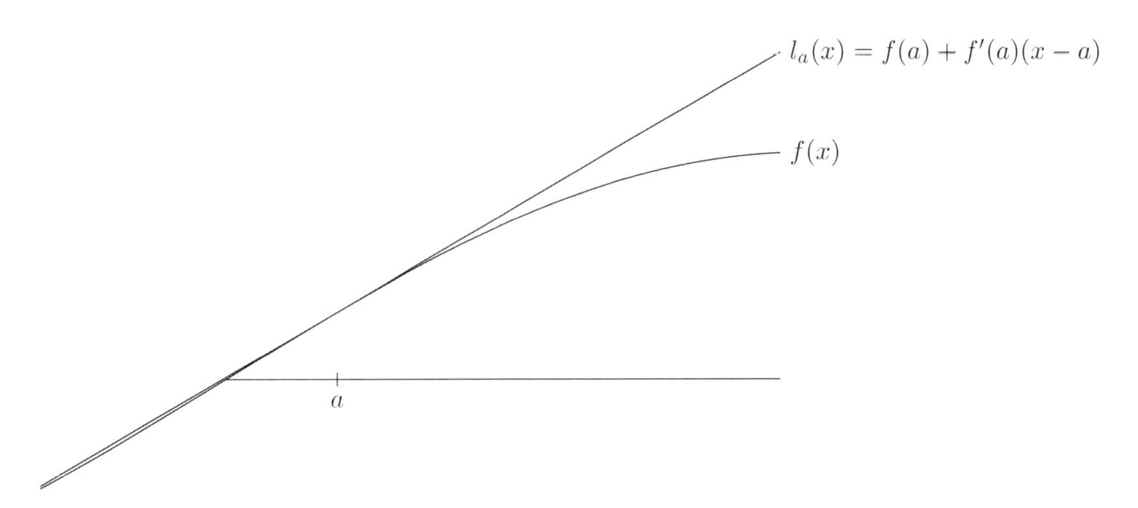

The tangent line, calculated by using the first derivative, provides the best linear approximation for the function f(x). The Taylor Series approximation uses higher-order derivatives to approximate the function with higher-degree polynomials. Figure A.2 shows an approximation of the function e^x by a degree 1, degree 2, and degree 3 polynomial.

Figure A.2
Taylor Series Approximation

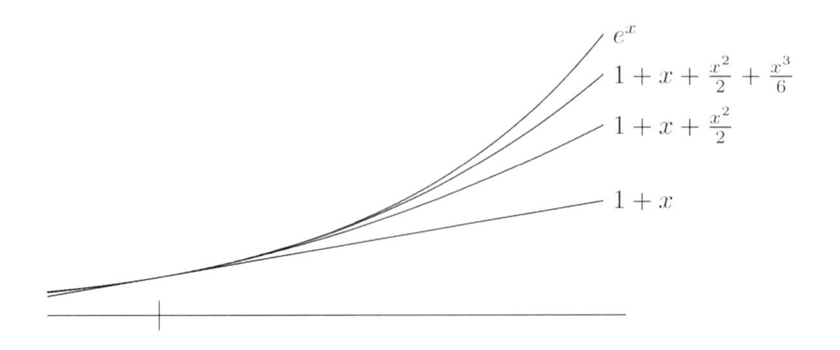

The tangent line to a function, $f(x)$, at a point a is the linear function that has the same first derivative as $f(x)$. The Taylor Series approximation is created by matching higher-order derivatives of the function $f(x)$.

The general Taylor Series approximation to a function a $f(x)$ at a point a is given by the formula

$$f(x)=\sum_{n=0}^{\infty}\frac{f^{(n)}(a)}{n!}(x-a)^n.$$

This sum can be broken into two pieces,

$$f(x)=\underbrace{\sum_{n=0}^{N}\frac{f^{(n)}(a)}{n!}(x-a)^n}_{P_N(x)}+\underbrace{\sum_{n=N+1}^{\infty}\frac{f^{(n)}(a)}{n!}(x-a)^n}_{E_N(x)}.$$

The first piece, $P_N(x)$, provides an approximation to the function $f(x)$ by a polynomial of degree N, while the second term gives the error of this approximation.

Taylor Series approximations are an extremely useful computational tool, and these approximations are widely used (e.g., in calculators and software systems) to calculate more complicated functions.

Figure A.2 gives the first three approximations for the exponential function $f(x)=e^x$ at the point 0. The general Taylor Series for the exponential function at $a=0$ is given by the sum

$$e^x=\sum_{n=0}^{\infty}\frac{x^n}{n!}.$$

Truncating this sum yields an approximation to the exponential function

$$e^x\approx\sum_{n=0}^{N}\frac{x^n}{n!}.$$

Using this approximation, an exponential function can be computed to arbitrary precision using only the operations of addition and multiplication.

In particular, the accuracy of this approximation—for some c between 0 and x—is given by

$$\left|e^x-\sum_{n=0}^{N}\frac{x^n}{n!}\right|=\left|\sum_{n=N+1}^{\infty}\frac{x^n}{n!}\right|=\frac{e^c}{(N+1)!}x^{N+1}.$$

Error Function

The standard normal distribution has pdf

$$f(x)=\frac{1}{\sqrt{2\pi}}e^{-x^2/2}.$$

This gives rise to a "Bell Curve," shown in Figure A.3.

Figure A.3
Standard Normal Distribution

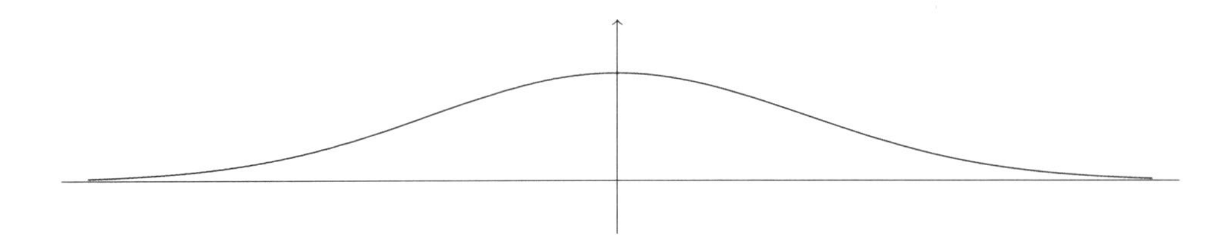

Working with normally distributed random variables requires finding the probability that they fall in a certain interval. For this, the cumulative distribution function is needed. For a random variable X with standard normal distribution, the cumulative distribution function of X gives the probability that $X < x$ is given by the formula

$$\frac{1}{\sqrt{2\pi}}\int_{-\infty}^{x} e^{-x^2/2}\,dx$$

Figure A.4
Cumulative Distribution Function

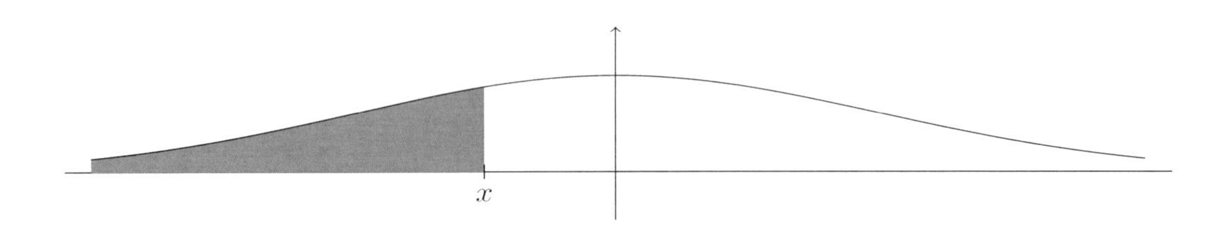

The cumulative distribution function measures the area under the normal distribution shown in blue in Figure A.4. It is standard to write the cumulative distribution function in the following manner:

$$\begin{aligned}\Pr[X < x] &= \frac{1}{\sqrt{2\pi}}\int_{-\infty}^{x} e^{-t^2/2}dt \\ &= \frac{1}{2} + \frac{1}{\sqrt{2\pi}}\int_{0}^{x} e^{-t^2/2}dt \\ &= \frac{1}{2} + \frac{1}{\sqrt{2\pi}}\sqrt{2}\int_{0}^{x/\sqrt{2}} e^{-t^2}\,dt \\ &= \frac{1}{2}\left[1 + \operatorname{erf}\left(\frac{x}{\sqrt{2}}\right)\right]\end{aligned}$$

Where $\operatorname{erf}(z) = \frac{2}{\sqrt{\pi}}\int_0^z e^{-t^2}\,dt$ is known as the *error function*. The preceding equations show that calculating the cumulative distribution function for a standard normal random variable can be reduced to calculating the error function. There is no closed-form expression for the error function, and values of the error function must be estimated numerically.

Obtaining an estimate for the error function relies on the Taylor Series expansion for the exponential function given by

$$e^t = \sum_{k=0}^{\infty}\frac{t^k}{k!} = 1 + t + \frac{t^2}{2} + \frac{t^3}{6} + \frac{t^4}{24} + \frac{t^5}{120} + \cdots$$

Combining the Taylor Series expansion for e^t with the definition of the error function yields

$$\begin{aligned}
\operatorname{erf}(z) &= \frac{2}{\sqrt{\pi}}\int_0^z e^{-t^2}\,dt \\
&= \frac{2}{\sqrt{\pi}}\int_0^z \sum_{k=0}^{\infty}\frac{(-t^2)^k}{k!}dt \\
&= \frac{2}{\sqrt{\pi}}\int_0^z \sum_{k=0}^{\infty}\frac{(-1)^k}{k!}t^{2k}dt \\
&= \frac{2}{\sqrt{\pi}}\sum_{k=0}^{\infty}\frac{(-1)^k}{k!}\int_0^z t^{2k}dt \\
&= \frac{2}{\sqrt{\pi}}\sum_{k=0}^{\infty}\frac{(-1)^k}{k!}\frac{z^{2k+1}}{2k+1} \\
&= \frac{2}{\sqrt{\pi}}\sum_{k=0}^{\infty}\frac{(-1)^k z^{2k+1}}{(2k+1)k!} \\
&= \frac{2}{\sqrt{\pi}}\left[z - \frac{z^3}{3} + \frac{z^5}{10} - \frac{z^7}{42} + \cdots\right]
\end{aligned}$$

The summation, $\operatorname{erf}(z) = \frac{2}{\sqrt{\pi}}\sum_{k=0}^{\infty}\frac{(-1)^k z^{2k+1}}{(2k+1)k!}$, provides an efficient means of evaluating the error function, using only the operations of addition and multiplication.

Estimating Integrals

The integral of a function $f(x)$ can be viewed geometrically as the area under the curve f. For example, in Figure A.5, the integral $\int_a^b f(x)dx$ represents the shaded area under the curve $f(x)$ between a and b.

Figure A.5
Definite Integral

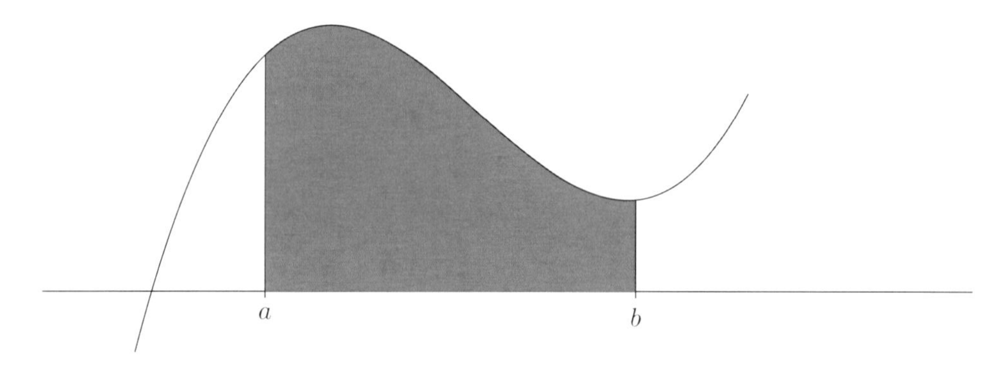

Integrals play an important role in working with probabilities. If $f(x)$ is a pdf of a random variable X, then $\int_a^b f(x)dx$ represents the probability that X falls between a and b.

Many integrals (e.g., $\int e^{x^2} dx$) do not have a closed-form expression, and must be estimated numerically. The simplest method for estimating an integral numerically is the Riemann Sum, named for mathematician Bernhard Riemann. To estimate the area under the curve f, the interval $[a,b]$ is broken into rectangles of width Δx, whose upper left corner lies on the curve $f(x)$ (see Figure A.6).

Figure A.6
Riemann Approximation

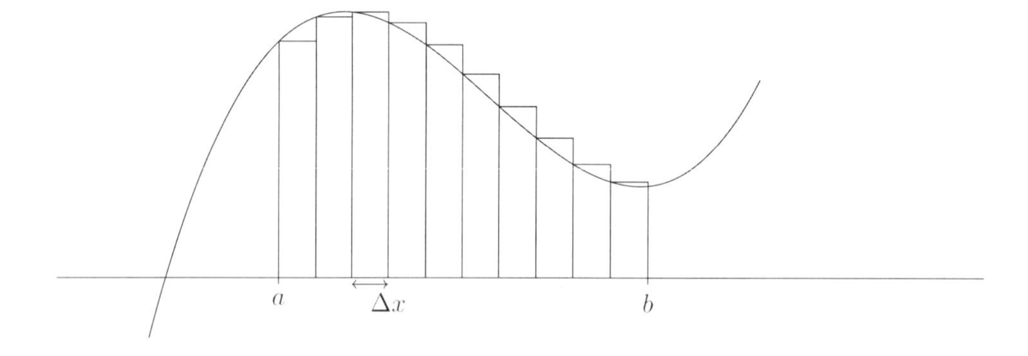

The height of the first rectangle is $f(a)$, and the width is Δx, so the area is $f(a)\Delta x$. The height of the second rectangle is $f(a+\Delta x)$, and the width is also Δx, so the area is $f(a+\Delta x)\Delta x$. Adding the area of all the rectangles in this way gives an estimate

$$\int_a^b f(x)dx \approx \sum_{n=0}^{N-1} f(a+n\Delta x)\Delta x.$$

As the number of rectangles, N, increases, the width Δx gets smaller since $\Delta x = \frac{b-a}{N}$. As the rectangles get narrower, the approximation becomes more accurate. The Riemann Sum provides a means of estimating an integral to an arbitrary degree of precision using only N evaluations of the function f and one multiplication. If the function f can be evaluated using only addition and multiplication operations, then this provides a means for evaluating the integral using only addition and multiplication operations.

Simpson's Rule

Where the Riemann Sum approximates an integral by a series of rectangles, a more accurate approximation can be obtained by approximating the integral by a series of arcs of parabola.

A parabola is given by an equation $y = ax^2 + bx + c$. Given three points $(x_0, y_0), (x_1, y_1), (x_2, y_2)$, it is possible to find the coefficients a, b, c so that the parabola passes through these points (see Figure A.7).

$$\int_a^b f(x)dx \approx \frac{\Delta x}{3}(f(a) + 4f(a+\Delta x) + 2f(a+2\Delta x) + 4f(a+3\Delta x) + 2f(a+4\Delta x) + \cdots + 4f(b-\Delta x) + f(b)).$$

Figure A.7
Area Under the Arc of a Parabola

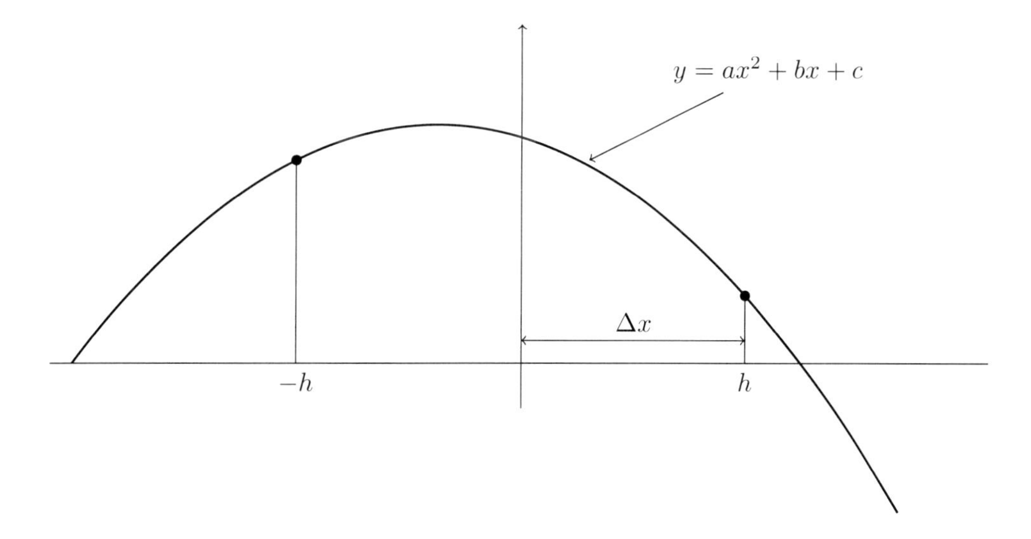

The area under the parabola is given by

$$\int_{-h}^{h} ax^2 + bx + cdx = \left(\frac{ax^3}{3} + \frac{bx^2}{2} + cx \right)\Bigg|_{-h}^{h}$$

$$= \frac{2ah^3}{3} + 2ch$$

$$= \frac{h}{3}(2ah^2 + 6c)$$

Since the three points, $(-h, y_0), (0, y_1), (h, y_2)$ are on the parabola, we have the equations

$$y_0 = ah^2 - bh + c$$

$$y_1 = c$$

$$y_2 = ah^2 + bh + c$$

Thus

$$\int_{-h}^{h} ax^2 + bx + cdx = \frac{h}{3}(2ah^2 + 6c)$$

$$= \frac{h}{3}((ah^2 - bh + c) + 4c + (ah^2 + bh + c))$$

$$= \frac{h}{3}(y_0 + 4y_1 + y_2)$$

$$= \frac{\Delta x}{3}(y_0 + 4y_1 + y_2).$$

To estimate the definite integral $\int_a^b f(x)dx$, break the interval $[a,b]$ into an even number, n, of subintervals of width $\Delta x = \frac{b-a}{n}$, i.e.,

$$a = x_0, \quad x_1 = a + \Delta x, \quad \ldots, \quad x_n = a + n\Delta x = b.$$

This gives

$$\int_a^b f(x)dx = \int_{x_0}^{x_2} f(x)dx + \int_{x_2}^{x_4} f(x)dx + \cdots + \int_{x_{n-2}}^{x_n} f(x)dx$$

$$\approx \frac{\Delta x}{3}\big(f(x_0) + 4f(x_1) + f(x_2)\big) + \cdots + \frac{\Delta x}{3}\big(f(x_{n-2}) + 4f(x_{n-1}) + f(x_n)\big)$$

$$= \frac{\Delta x}{3}(f(x_0) + 4f(x_1) + 2f(x_2) + 4f(x_3) + \cdots + 4f(x_{n-1}) + f(x_n))$$

Separating the even and odd terms in the sum yields the approximation

$$\int_a^b f(x)dx \approx \frac{\Delta x}{3}(f(x_0) + f(x_n)) + \frac{4\Delta x}{3}\sum_{i=1}^{n/2} f(x_{2i-1}) + \frac{2\Delta x}{3}\sum_{i=1}^{n/2} f(x_{2i}).$$

Bibliography

Adida, Ben, "Helios: Web-based Open-Audit Voting," in *Proceedings of the 17th USENIX Security Symposium*, San Jose, Calif., July 28–August 1, 2008, pp. 335–348.

Alfano, Salvatore, "A Numerical Implementation of Spherical Object Collision Probability," *Journal of the Astronautical Sciences*, Vol. 53, No. 1, 2001, pp. 103–109.

Baiocchi, Dave, and William Welser IV, *Confronting Space Debris: Strategies and Warnings from Comparable Examples Including Deepwater Horizon*, Santa Monica, Calif.: RAND Corporation, MG-1042-DARPA, 2010. As of November 12, 2012: http://www.rand.org/pubs/monographs/MG1042.html

Ben-David, Assaf, Noam Nisan, and Benny Pinkas, "FairplayMP—A System for Secure Multi-Party Computation," in *CCS '08 Proceedings of the 15th ACM Conference on Computer and Communications Security*, New York: ACM, 2008, pp. 257–266.

Ben-Or, Michael, Shafi Goldwasser, and Avi Wigderson, "Completeness Theorems for Non-Cryptographic Fault-Tolerant Distributed Computation," in *Proceedings of the 20th Annual ACM Symposium on Theory of Computing*, Chicago, Ill., 1988, pp. 1–10.

Bogdanov, D., S. Laur, and J. Willemson, "Sharemind: A Framework for Fast Privacy-Preserving Computations," in *Proceedings of the 13th European Symposium on Research in Computer Security: Computer Security*, ser. ESORICS '08, Vol. 5283, Berlin, Heidelberg: Springer-Verlag, 2008, pp. 192–206.

Bogetoft, Peter, Dan Lund Christensen, Ivan Damgård, Martin Geisler, Thomas Jakobsen, Mikkel Krøigaard, Janus Dam Nielsen, Jesper Buus Nielsen, Kurt Nielsen, Jakob Pagter, Michael Schwartzbach, and Tomas Toft, *Multiparty Computation Goes Live*, Report 2008/068, IACR Cryptology ePrint Archive, 2008. As of November 12, 2012:
http://eprint.iacr.org/2008/068.pdf

Center for Space Standards and Innovation, Satellite Orbital Conjunction Reports Assessing Threatening Encounters in Space (SOCRATES), online. As of November 12, 2012:
http://www.celestrak.com/SOCRATES/

———, *Chinese ASAT Test*, online. As of August 13, 2013:
http://www.centerforspace.com/asat/

Chaum, David, Claude Crépeau, and Ivan Damgård, "Multiparty Unconditionally Secure Protocols," *Proceedings of the 20th Annual ACM Symposium on Theory of Computing*, Chicago, Ill., 1988, pp. 11–19.

Chevillard, Sylvain, and Nathalie Revol, "Computation of the Error Function erf in Arbitrary Precision with Correct Rounding," *RNC 8, the 8th Conference on Real Numbers and Computers,* Santiago de Compostela, Spain, 2008, pp. 27–36.

Choi, Seung Geol, Kyung-Wook Hwang, Jonathan Katz, Tal Malkin, and Dan Rubenstein, *Secure Multi-Party Computation of Boolean Circuits with Applications to Privacy in On-Line Marketplaces*, IACR Cryptology ePrint Archive, Report 2011/257, 2011. As of November 12, 2012: http://eprint.iacr.org/2011/257

Chow, Tiffany, "SSA Sharing Program," Secure World Foundation Issue Brief, October 5, 2010.

Damgård, Ivan, Martin Geisler, Mikkel Krøigaard, and Jesper B. Nielsen, "Asynchronous Multiparty Computation: Theory and Implementation," *Public Key Cryptography - PKC 2009, 12th International Conference on Practice and Theory in Public Key Cryptography, Irvine, CA, USA, March 18–20, 2009, Proceedings*, Springer, 2009, pp. 160–179.

Foust, Jeff, "A New Eye in the Sky to Keep an Eye on the Sky," *The Space Review*, May 10, 2010.

Franklin, Matthew, and Moti Yung, "Varieties of Secure Distributed Computing," in *Proceedings of Sequences II, Methods in Communications, Security and Computer Science*, 1996.

Goldreich, Oded, *Foundations of Cryptography,* Volume II, Cambridge University Press, 2004.

Goldreich, Oded, Sylvio Micali, and Avi Wigderson, "How to Play ANY Mental Game," *STOC 1987: Proceedings of the Nineteenth Annual ACM Symposium on Theory of Computing*, New York: ACM, January 1987, pp. 218–229.

Hall, Robert, Salvatore Alfano, and Alan Ocampo, "Advances in Satellite Conjunction Analysis," *Proceedings of the Advanced Maui Optical and Space Surveillance Technologies Conference*, 2010.

Hashemian, Reza, "A New Multiplier Using Wallace Structure and Carry Select Adder with Pipelining," *ISCAS '02 Conference Proceedings*, 2002.

Hazay, Carmit, and Yehuda Lindell, *Efficient Secure Two-Party Protocols: Techniques and Constructions*, Springer, 2010.

Huang, Y., C. H. Shen, D. Evans, J. Katz, and A. Shelat, "Efficient Secure Computation with Garbled Circuits," *Proceedings of the 7th International Conference on Information Systems Security*, ser. ICISS'11, Berlin, Heidelberg: Springer-Verlag, 2011, pp. 28–48.

Institut français des relations internationals, “Assessing the Current Dynamics of Space Security,” presented at SWF-Ifri workshop, Paris, June 18–19, 2009. As of November 12, 2012:
http://swfound.org/media/1878/spacesecurity-ifri-rw-2009.pdf

Ishai, Yuval, Joe Kilian, Kobbi Nissim, and Erez Petrank, "Extending Oblivious Transfers Efficiently," *CRYPTO*, 2003, pp. 145–161.

Kelso, T. S., “How the Space Data Center is Improving Safety of Space Operations,” presented at the 13th Advanced Maui Optical and Space Surveillance (AMOS) Technologies Conference, Maui, Hawaii, September 16, 2010.

Kelso, T. S., David A. Vallado, Joseph Chan, and Bjorn Buckwalter, “Improved Conjunction Analysis via Collaborative Space Situational Awareness,” presented at the 9th Advanced Maui Optical and Space Surveillance (AMOS) Technologies Conference, Maui, Hawaii, September 19, 2008.

Kolesnikov, V. and T. Schneider, “Improved Garbled Circuit: Free XOR Gates and Applications,” *Proceedings of the 35th International Colloquium on Automata, Languages and Programming, Part II*, ICALP '08, Berlin, Heidelberg: Springer-Verlag, 2008, pp. 486–498.

Kreuter, B, Abhi Shelat, and Chi-Hao Shen, "Billion-Gate Secure Computation with Malicious Adversaries," *Proceedings of the 21st USENIX Conference on Security Symposium*, ser. Security'12, Berkeley, Calif.: USENIX Association, 2012, p. 14.

Lindell, Yehuda, and Benny Pinkas, *A Proof of Security of Yao's Protocol for Two-Party Computation,* ePrint 2004/175, 2004.

———, “An Efficient Protocol for Secure Two-Party Computation in the Presence of Malicious Adversaries,” in M. Naor, ed., *EUROCRYPT 2007*, pp. 52–78.

———, “Secure Multiparty Computation for Privacy-Preserving Data Mining,” *The Journal of Privacy and Confidentiality*, Vol. 1, No. 1, 2009, pp. 59–98.

———, “Secure Two-Party Computation via Cut-and-Choose Oblivious Transfer,” in Y. Ishai, ed., *Theory of Cryptography*, Vol. 6597 of *Lecture Notes in Computer Science*, Berlin, Heidelberg: Springer, 2011, pp. 329–346.

Malka, Lior, "VMCrypt: Modular Software Architecture for Scalable Secure Computation," *Proceedings of the 18th ACM Conference on Computer and Communications Security*, ser. CCS '11, New York: ACM, 2011, pp. 715–724.

Malkhi, Dahlia, Noan Nisan, Benny Pinkas, and Yaron Sella, "Fairplay - A Secure Two-Party Computation System," *USENIX Security Symposium '04*, 2004.

Mardziel, P., M. Hicks, J. Katz, and M. Srivatsa, "Knowledge-Oriented Secure Multiparty Computation," *Proceedings of the 7th Workshop on Programming Languages and Analysis for Security*, ser. PLAS '12, New York: ACM, 2012.

Nielsen, Jesper Buus, Peter Sebastian Nordholt, Claudio Orlandi, and Sai Sheshank Burra, *A New Approach to Practical Active-Secure Two-Party Computation*, IACR Cryptology ePrint Archive, Report 2011/091, 2011. As of November 12, 2012: http://eprint.iacr.org/2011/091

Pinkas, B., T. Schneider, N. P. Smart, and S. C. Williams "Secure Two-Party Computation Is Practical," *Proceedings of the 15th International Conference on the Theory and Application of Cryptology and Information Security: Advances in Cryptology*, ASIACRYPT '09, Berlin, Heidelberg: Springer-Verlag, 2009, pp. 250–267.

Shamir, Adi, "How to Share a Secret," *Communications of the ACM*, Vol. 22, No. 11, November 1979, pp. 612–613.

Sharemind, home page, January 2012. As of November 12, 2012: http://sharemind.cyber.ee/

Space Data Association, "Space Data Center Attains Full Operational Capability Status," press release, September 9, 2011. As of November 12, 2012: http://www.space-data.org/sda/wp-content/uploads/downloads/2011/10/20110919_SDA_release_FOC.pdf

Université Catholique de Louvain, "Un Système de Vote Inédit à Grande Échelle," March 5, 2009. As of November 12, 2012: http://www.uclouvain.be/270428.html

VIFF, the Virtual Ideal Functionality Framework, home page, January 2012. As of November 12, 2012: http://viff.dk/

Weeden, Brian, Paul Cefola, and Jaganathan Sankaran, "Global Space Situational Awareness Sensors," presented at the 11th Advanced Maui Optical and Space Surveillance (AMOS) Technologies Conference, Maui, Hawaii, September 16, 2010.

The White House, *National Space Policy of the United States of America,* Washington, D.C., June 28, 2010. As of November 12, 2012: http://www.whitehouse.gov/sites/default/files/national_space_policy_6-28-10.pdf

Yao, Andrew C., "Protocols for Secure Computations," *23rd Annual Symposium on Foundations of Computer Science (FOCS 1982)*, Chicago, Ill., November 3–5, 1982, pp. 160–164.

Yao, Andrew C., "How to Generate and Exchange Secrets," *27th Annual Symposium on Foundations of Computer Science (FOCS 1986)*, Toronto, October 27–29, 1986, pp. 162–167.